Engraved by W.T.Fry.

SIR ISAAC NEWTON.

OPTICKS

OR

A Treatise of the Reflections, Refractions, Inflections & Colours of Light

SIR ISAAC NEWTON

BASED ON THE FOURTH EDITION LONDON, 1730

With a Foreword by
ALBERT EINSTEIN
An Introduction by
SIR EDMUND WHITTAKER
A Preface by
I. BERNARD COHEN
*And an Analytical Table of Contents
prepared by*
DUANE H. D. ROLLER

DOVER PUBLICATIONS, INC.
NEW YORK

This Dover edition, first published in 1952, is an unabridged and unaltered republication of the work originally published by G. Bell and Sons, Ltd., in 1931. New to this edition are a Preface by I. Bernard Cohen and an Analytical Table of Contents prepared by Duane H. D. Roller.
This work is reprinted by special arrangement with G. Bell and Sons, Ltd.

Standard Book Number: 486-60205-2
Library of Congress Catalog Card Number: 52-12165

Manufactured in the United States of America
Dover Publications, Inc.
180 Varick Street
New York, N. Y. 10014

This edition of Newton's *Opticks*
is dedicated to
MARJORIE HOPE NICOLSON,
who has uncovered the influence of the
Opticks on 18th-century imagination.

CONTENTS

PREFACE

GREAT CREATIONS—whether of science or of art—can never be viewed dispassionately. The *Opticks,* like any other scientific masterpiece, is a difficult book to view objectively; first, because of the unique place of its author, Isaac Newton, in the history of science, and, second, because of the doctrine it contains. One of the most readable of all the great books in the history of physical science, the *Opticks* remained out of print for a century and a half, until about two decades ago, while the *Principia* was constantly being reprinted. One of the reasons for this neglect was that the *Opticks* was out of harmony with the ideas of 19th-century physics. The burden of this book was an exposition of the "wrong" (i.e., corpuscular) theory of light,—even though it also contained many of the basic principles of the "correct" (i.e., wave) theory. Not only had Newton erred in his choice of the corpuscular theory, but also he apparently had found no insuperable difficulty in simultaneously embracing features of two opposing theories. It was not just that

Newton couldn't make up his mind between them; that could be easily forgiven. Rather, by adopting a combination of the two theories at once, he had violated one of the major canons of 19th-century physics, which held that whenever there are two conflicting theories, a crucial experiment must always decide uniquely in favor of one or the other.

Every age finds a particular sympathy for certain masterpieces of science and of art. Our present age lends a particularly appreciative eye and ear to the paintings of Lucas Cranach the elder and Jheronimus Bosch, and the poetry of John Donne and William Blake. Since our reading in the past great works of science is conditioned by the science of our own time, our interpretations and evaluations are as different from those of the last century as our tastes in poetry and art. We esteem the *Principia* as much as the Victorians did, but we know its limitations and cannot help but read it in the light of the theory of relativity. Today our point of view is influenced by the theory of photons and matter waves, or the more general principle of complementarity of Niels Bohr; and we may read with a new interest Newton's ideas on the interaction of light and matter or his explanation of the corpuscular and undulatory aspects of light. It is not surprising, therefore, that the last decades have witnessed a revival of Newton's *Opticks*, finally reprinted in 1931 by Messrs. G. Bell and

Sons of London with a foreword by Albert
Einstein and introduction by E. T. (now Sir
Edmund) Whittaker, a new printing of which
now makes this work again available thanks to
the enterprise of Mr. Hayward Cirker of the
Dover Publications. Sir Edmund's introduction
delineates the effect on the reputation of the
Opticks of the vicissitudes of the corpuscular
theory of light which Newton espoused, and the
general feeling during the latter 19th century
that since the wave theory of light was the only
true explanation of optical processes, the *Opticks*
was a work of interest chiefly to the historian
rather than the scientist, exhibiting, despite its
brilliant exposition, an unhappy example of how
wrong a great man might be.

So great, however, was Newton's fame among
men of science that a number of writers on optics, especially among the British, took care to
inform their readers that Newton's corpuscular
theory, while clearly incorrect, was nevertheless
a very ingenious creation and had been fully
able to explain all of the facts about light known
in Newton's day. In other words, this theory
was not wholly relegated to the realms of the
antique and the curious but was rather presented to the reader with an apology and a discussion of the 17th-century situation in physics.
The book itself makes plain that Newton's
theory was in fact adequate to the phenomena
it attempted to explain, and it was in many re-

spects better than the rival theory of Huygens.
Yet the sympathetic explanation of Newton's
approach to optics, in terms of the physics of
his day, is conspicuous in such books because the
authors of scientific treatises in the 19th century
were not usually quite so generous to their predecessors when it came to "erroneous" theories
of the past. Their attitude is all the more remarkable when we see that they usually felt
bound to add that Newton's choice of the
"wrong" theory of light had been a serious im-
pediment to the search for the "correct" theory.

Thus, in a book on optics oft reprinted during the 19th century, Dionysius Lardner wrote
of the corpuscular theory of light: ". . . probably, from veneration of his [Newton's] authority, English philosophers, until recently,
have very generally given the preference to that
theory."[1] And as late as 1909 Sir Arthur Schuster, in a general treatise on optics, still felt it
incumbent upon him to apologise for the apparent failure of the greatest of British men of
science and wrote: "While there is no doubt that
Newton's great authority kept back the progress of the undulatory theory for more than a
century, this is more than compensated by the
fact that the science of Optics owes the scientific
foundation of its experimental investigation in
great part to him."[2]

The revival of the wave theory of light, in
opposition to the corpuscular theory advocated

by Newton, was due largely to the labors of Dr. Thomas Young who, in a series of three papers published in 1802-04 in the *Philosophical Transactions* of the Royal Society[3], added a new fundamental principle, the so-called principle of interference, and established the wave theory on a basis that has from then until now remained unassailable[4]. Despite Young's insistence in all his papers that his own work derived from the experiments and suppositions of Newton, and his introduction of many quotations from Newton to show that he was modifying rather than destroying the Newtonian doctrine, Young was attacked mercilessly. The most important antagonist appears to have been Lord Brougham, in all probability the author of two anonymous discussions of Young's work in the *Edinburgh Review*. Brougham was particularly incensed at the suggestion of Newton's fallibility and he accused Young of having insinuated "that Sir Isaac Newton was but a sorry philosopher."[5] Had not Fresnel and Arago in France become interested in the work of Young, it seems probable that the influence of the great name of Newton would effectively have blocked any pursuit of Young's ideas—at least in England—and any further development of the wave theory of light. This situation is not unlike that in mathematics during the 18th century, when a blind adherence to the Newtonian algorithm and a complete rejection of the Leibnizian or Continental methods seem to have deadened the sensibilities of British mathematicians and to have produced an era of almost complete sterility with regard to progress in the calculus.[6]

The comparison of works of science and works of art is especially relevant to our understanding of the history of the reputation of the *Opticks*. Having known Picasso and Dali, Cranach and Bosch do not seem as strange to us as they would have to a Victorian spectator. Similarly, having known Planck, Einstein, Bohr, de Broglie, Schrödinger and Heisenberg, Newton's suggestions concerning the interaction of particles of light and particles of matter, and his adumbration of a corpuscular theory of light that also embraced undulatory behaviour, are not to us—as they were to the Victorian physicists—a puzzling set of ideas, seemingly devoid of any physical meaning and, therefore, outside the limits of comprehensibility. One of the great classic treatises on light of the late 19th century was written by Thomas Preston, Professor of Natural Philosophy in University College, Dublin. First printed in 1890, it still served as a principal textbook for this subject in the '30's (in the fifth edition published in 1928). It was notable for its magnificent expository style and the clarity of its explanations, and also for the valuable historical discussions and long extracts from Newton's *Opticks;* indeed, many students of physics, including the writer, learned of New-

ton's *Opticks* for the first time through reading
the selections printed in Preston's *Theory of
Light*. One group of extracts was intended to
present Newton's "theory in his own words" and
to "show how much more closely than is generally supposed it resembles the undulatory
theory now accepted." Preston wanted his readers to understand that Newton's position was, if
not exactly that of the late 19th century, at least
very much like it. By "suitably framing his
fundamental postulates," Preston wrote, "an
ingenious exponent of the emission [or corpus-
cular] theory . . . might fairly meet all the objections that have been raised against it." Yet,
when these "necessary postulates" are intro-
duced, the corpuscles become endowed "with the
periodic characteristics of a wave motion, and
. . . the corpuscles themselves may be elimi-
nated. . . ."[7] Thus Preston's attitude was to
"save the theory" by eliminating its fundamental
corpuscular character. Writing long before the
rise of quantum mechanics, Preston could not
find any virtue in a theory of light which simultaneously embraced corpuscular and undulatory
aspects and he therefore destroyed the former
in order to preserve the latter.

Yet if Newton's *Opticks* appears to contain
ideas that seem like anticipations of our present
concepts, we must be very careful not to make
too much of it. For, if we were to praise Newton today, simply because his statements about

the dual properties of light—corpuscular and undulatory—and the interaction of light and matter are in some ways like those of the 20th century, we are in a sense adopting a procedure of as little worth as did those who disparaged Newton 75 years ago because his optical theory did not quite meet the requirements of late 19th-century physics. However great Newton's insight may have been—and it was almost incredibly penetrating—it was hardly sufficient to give him a prevision of quantum theory; nor to enable him to see two centuries ahead to the science of spectroscopy built upon the foundation of his investigations of the prismatic spectrum of the sun; nor to allow him a guess as to what would be the problems of black-body radiation and the failure of the theorem of the equipartition of energy, the very data for which were not to be discovered until more than 150 years after his death. Nor could Newton have had even the faintest glimmering of the photo-electric effect and the theory of electrons, since in Newton's day the subject of electricity had not yet attained the status of a separate branch of science. Even the distinction between conductors and non-conductors had not yet been made; nor had the two kinds of electric charge (vitreous and resinous) been yet discovered; and Franklin, who produced in middle age the first unitary theory of electrical action, was not born until two years after the *Opticks* had been published.

We may find an appeal in the *Opticks* because we know the theory of photons and quantum mechanics, but we must also keep in mind that the physics of the 20th century was developed from the brute facts of experiment of the late 19th century and the early 20th, and in reaction to the inadequacy of the then-current theories which could not account for the incontrovertible observed data. The physics of the 20th century derives directly from the physics of the 19th century, not from a conscious return to the physics of the 18th century and Newton's *Opticks*. But the physics of the 20th century does in a definite sense derive from that of the 18th, since the latter produced the physics of the 19th century just as the physics of the 19th century, in turn, produced that of the 20th. In terms of the physics of our own time, then, the importance of the *Opticks* does not lie in any possible kind of prevision of the theory of photons, but rather in the effect it had on the physicists in its own day and on the generations immediately following its publication. In making an evaluation of the *Opticks,* therefore, we must choose between (1) the historical or (2) the antiquarian approach to the development of science—between the historian's evaluation of Newton's achievement in terms of the living creation and its influence on the scientists in the century following the publication of his results, or the antiquarian's sifting of the *disjecta membra* of the *Opticks*

(often out of context) for an occasional "pre-
cursorship" of one or another 20th-century
physical concept.

My own interest in the *Opticks* was aroused
in the course of extended research into the
theories of electricity as developed in the 18th
century, as a part of a larger study on the
growth in the 18th century of concepts crucial
in 19th-century physics, such as charge, field, potential, force, action-at-a distance, atom, energy,
etc.[8] I soon found that of the many references
to Newton in 18th-century electrical writings
only a very small number were to the *Principia,*
the greater part by far were to the *Opticks*. This
was true not alone of the electrical writings but
also in other fields of experimental enquiry.

As an example of the influence of the *Opticks,*
we may look at the *Vegetable Staticks* of Stephen
Hales, a work of the highest rank among experimental treatises, and one that has earned for its
author the title of father of plant physiology.
This book provides a splendid example of the
application of quantitative methods to biology
in its account of precise measurements of leaf
growth, root pressure, and kindred subjects, and
its quantitative studies on air and gases, just as in
the companion volume, *Haemostaticks,* Hales
initiated measurements of the blood pressure in
animals. The *Vegetable Staticks* bears on the
verso of the title page Newton's imprimatur as
President of the Royal Society: "Feb. 16,

1726/7. *Imprimatur* Isaac Newton. *Pr. Reg. Soc.*" The first mention of Newton occurs in the preface, where Hales notes how "it appears by many chymico-statical Experiments, that there is diffused thro' all natural, mutually attracting bodies, a large proportion of particles, which, as the first Author of this important discovery, Sir Isaac Newton, observes, are capable of being thrown off from dense bodies by heat or fermentation into a vigorously elastick and repelling state . . . ", a reference to the *Opticks* (see, below, Qu. 30 and Qu. 31). In addition to a discussion of "attraction: that universal principle which is so operative in all the very different works of nature,"[9] without mentioning Newton's name, there are by actual count 17 places in the book where Newton's name occurs. Of these, 15 are either quotations from the *Opticks* or references to the *Opticks,* and neither of the remaining two are concerned with the *Principia:* one discusses Newton's mode of calibrating thermometers (described in the *Phil. Trans. Roy. Soc.*[10]) and the other is a quotation containing Newton's theory about the dissolution of metals arising from the attractive force exerted by the "particles of acids" (taken from the introduction to vol. 2 of John Harris' *Lexicon Technicum*[11]).

In order to understand the extraordinary appeal that the *Opticks* had in the 18th century, we must compare it to the *Principia*[12]—in scientific, philosophic, and speculative content; literary style; and the approach of the author to the subject. On such a comparison, an important difference between the two books is immediately apparent. The *Opticks* invites and holds the attention of the non-specialist reader while the other, the *Principia,* is as austere and forbidding to the non-specialist as it can possibly be. Of course, the general reader of the *Opticks* would be more interested in the final section of "Queries" than in the rest of the work, just as the general reader of the *Principia* would be drawn to the General Scholium at the end of Book Three; but whereas in the *Opticks* such a reader could enjoy almost 70 pages, in the *Principia* there would be but four. The latter would discuss for him the mechanism of universal gravitation and give him a hint of the direction of Newton's thinking about this important problem; but the former would allow the reader to roam, with great Newton as his guide, through the major unresolved problems of science and even the relation of the whole world of nature to Him who had created it.

Wholly apart from the general reader, the scientist who was not well trained in mathematics could make little headway in the *Principia*. Not only was this masterpiece written in an austere mathematical style, consisting largely of definitions, theorems, lemmas, scholia, and demonstrations, but it was in a definite sense written in an archaic mathematical language. Newton did not consistently apply his own discovery of the calculus, but preferred to use the geometrical style of Apollonios and Euclid—whose works he recommended, along with others, to the theologian William Bentley who wished to present the work of Newton in a popular way as proof of the wisdom of the Creator of the universe. Bentley had first written to the Scotch mathematician John Craigie[13] and appears to have been so alarmed by the number of mathematical authors recommended by Craigie as being necessary to understand the *Principia,* that he then applied to Newton himself for advice. The latter's list is formidable enough; a non-mathematician would have to be in earnest indeed to undertake the preparation which Newton deemed necessary for reading the *Principia* even with the limited objective of a "first perusal," for which Newton advised that "it's enough if you understand the Propositions with some of the Demonstrations which are easier than the rest."[14]

The famous philosopher John Locke was more sensible than Bentley and freely admitted that his mathematics would never be equal to reading the great book. He was satisfied with an examination of the reasoning behind the propositions and corollaries to be drawn from them. To be sure that all was well, he inquired of the Dutch physicist Christiaan Huygens whether

the mathematics were sound, and once assured
that this was the case he was content with the
physical principles, the doctrine the book ex-
pounded, without bothering about the proofs
and details of the text itself. We have an account of Locke's procedure written by Newton's
disciple and friend, the Rev. J. T. Desaguliers,
who informs us that he was told the story "several times by Sir *Isaac Newton* himself":

> But to return to the *Newtonian Philoso-
> phy:* Tho' its Truth is supported by Mathe-
> maticks, yet its Physical Discoveries may
> be communicated without. The great Mr.
> *Locke* was the first who became a *New-
> tonian Philosopher* without the help of
> Geometry; for having asked Mr. *Huygens,*
> whether all the mathematical *Propositions*
> in Sir *Isaac's Principia* were true, and
> being told he might depend upon their
> Certainty; he took them for granted, and
> carefully examined the Reasonings and
> *Corollaries* drawn from them, became
> Master of all the Physics, and was fully
> convinc'd of the great Discoveries con-
> tained in that Book.[15]

Newton himself had given a warrant for Locke's
procedure in the opening lines of the third book
of the *Principia*, where he wrote that in the two
preceding books he had "laid down the principles of philosophy; principles not philosophi-

cal but mathematical," that these "principles are
the laws and conditions of certain motions, and
powers of forces"; lest they "should have appeared of themselves dry and barren, I have
illustrated them here and there with some philosophical scholiums. . . ."

Desaguliers contrasts Locke's reading of the
Principia with the way in which "he read the
Opticks with pleasure, acquainting himself with
every thing in them that was not merely mathematical."[16] The merely mathematical section
consisted of "Two Treatises of the Species and
Magnitude of Curvilinear Figures," which Newton omitted after the first edition, since they
were in no way connected with the text of the
Opticks.

Another distinction between the *Opticks* and
the *Principia,* apart from the mathematical difficulty of the latter, is that the *Opticks* was written
in English while the *Principia* was written in
Latin. While this did not make as much difference in the 17th century as it would today, there
is no question but what the austere Latin of the
Principia was as characteristic of its essentially
mathematical form as the gentle English of the
Opticks was characteristic of the intimate style
of that work[17]. For in the *Opticks* Newton did
not adopt the motto to be found in the *Principia*
—*Hypotheses non fingo;* I frame no hypotheses
—but, so to speak, let himself go, allowing his

imagination full reign and by far exceeding the
bounds of experimental evidence.

It should, of course, be borne in mind that
Newton's phrase *Hypotheses non fingo* was applied by him to the nature of the gravitational
attraction and was never a guiding principle in
his work. It is equally clear, however, that many
of the readers of the *Principia* tended to think
of this motto as characteristic of the book. Thus
Roger Cotes, who superintended the preparation of the second edition of the *Principia* under
Newton's direction and who wrote a preface to
it, begged Newton to revise "the last Sheet of
your Book which is not yet printed off," since
he felt he could not "undertake to answer any
one who should assert that You do *Hypothesim
fingere,* [since] I think You seem tacitly to make
this supposition that the Attractive force resides
in the Central Body."[18] In a letter written in
1672 by Newton to Henry Oldenberg, Secretary
of the Royal Society, in response to an objection that had been raised to his first publication
on optics, Newton discussed the function of
hypotheses at length:

> For the best and safest method of phi-
> losophizing seems to be, first diligently to
> investigate the properties of things and
> establish them by experiment, and then to
> seek hypotheses to explain them. For hy-
> potheses ought to be fitted merely to ex-

plain the properties of things and not attempt to predetermine them except in so far as they can be an aid to experiments. If any one offers conjectures about the truth of things from the mere possibility of hypotheses, I do not see how anything certain can be determined in any science; for it is always possible to contrive hypotheses, one after another, which are found rich in new tribulations. Wherefore I judged that one should abstain from considering hypotheses as from a fallacious argument, and that the force of their opposition must be removed, that one may arrive at a maturer and more general explanation.[19]

The first book of the *Opticks* deals with the reflection and refraction of light, the formation of images, the production of spectra by prisms, the properties of colored light and the composition of white light and its dispersion. Based on definitions and axioms, and embodying a wealth of experimental data, this first book had, according to Newton, the "Design . . . not to explain the Properties of Light by Hypotheses, but to propose and prove them by Reason and Experiments." The second book, devoted largely to the production of colors in what we would call interference phenomena, contains no such declaration, and it is here that Newton introduces the notion of "fits" of easy transmission and easy reflection, and kindred concepts not derived by induction from experiments. And although Newton points out (p. 280) that on the score of fits of easy transmission and easy reflection: "What kind of action or disposition this is; Whether it consists in a circulating or a vibrating motion of the Ray, or of the Medium, or something else, I do not here enquire," he adds:

> Those that are averse from assenting to any new Discoveries, but such as they can explain by an Hypothesis, may for the present suppose, that as Stones by falling upon Water put the Water into an undulating Motion, and all Bodies by percussion excite vibrations in the Air; so the Rays of Light, by impinging on any refracting or reflecting Surface, excite vibrations in the refracting or reflecting Medium or Substance, and by exciting them agitate the solid parts of the refracting or reflecting Body, and by agitating them cause the Body to grow warm or hot; that the vibrations thus excited . . . move faster than the Rays so as to overtake them; . . . and, by consequence, that every Ray is successively disposed to be easily reflected, or easily transmitted, by every vibration which overtakes it. But whether this Hypothesis be true or false I do not here consider. I content my self with the bare Discovery,

that the Rays of Light are by some cause
or other alternately disposed to be reflected or refracted for many vicissitudes.

The second book thus admits hypotheses, although without any consideration of their truth
or falsity. In the third (and last) book, the
opening section deals with Newton's experiments on diffraction, followed by the famous
Queries in which, as we shall see, Newton intro-
duced a variety of "hypotheses"—not alone on
light, but on a great many subjects of physics
and philosophy, as if in his final work he had
emptied his mind of the conjectures he had ac-
cumulated in a life-time of scientific activity.
Clearly, *Hypotheses non fingo* could not be applied to the *Opticks*. And it is, in a very real
sense, the progressively conjectural character of
this book that makes it so interesting to read. As
Albert Einstein saw so clearly when he wrote
his admirable Foreword, this book "alone can
afford us the enjoyment of a look at the personal
activity of this unique man."

In its own day, the *Opticks* aroused interest
in a way that was directly related to the *Principia*. Not only did the reputation of the *Principia* create a ready market for a more readable
book by its author, but in the *Principia* Newton
had raised important philosophical questions
which he discussed at greater length in the
Opticks, in the Queries at the end of Book

Three, and which Newton mentioned—but only in passing—in the famous General Scholium to the third book of the *Principia,* addressed to the nature of the gravitational attraction between bodies, and in which the phrase *Hypotheses non fingo* appeared.

Newton had shown that celestial and terrestrial motions were in accordance with a law of universal gravitation in which the attraction between any two bodies in the universe depends only on their masses and (inversely) on the square of the distance between them. This led to an attribution to Newton of ideas that he abhorred. One was that since the gravitational attraction is a function of the masses of bodies irrespective of any other properties save their separation in space, this attraction arises simply from the existence of matter. This materalist position was castigated by Newton in a letter to Bentley in which he said: "You sometimes speak of gravity as essential and inherent to matter. Pray, do not ascribe that notion to me; for the cause of gravity is what I do not pretend to know. . . ." And in another letter to Bentley, he amplified his position: "It is inconceivable, that inanimate brute matter should, without the mediation of something else, which is not material, operate upon and affect other matter without mutual contact. . . ."[20]

Another point of argument arose in a letter written by Leibniz which had been published in an English translation. Cotes wrote to Newton of "some prejudices which have been industriously laid against" the *Principia*, "As that it deserts Mechanical causes, is built upon Miracles, & recurrs to Occult qualitys." Newton would find "a very extraordinary Letter of Mr Leibnitz to Mr Hartsoeker which will confirm what I have said," Cotes continued, in "a Weekly Paper called *Memoires of Literature* & sold by Ann Baldwin in Warwick-Lane."[21]

In the preface which he wrote to the second edition of the *Principia*, Cotes replied to Leibniz—although without mentioning his name; ". . . twere better to neglect him," he had written to Newton.[22] Cotes also discussed the general nature of gravitation and forces acting at a distance. For this second edition, Newton wrote the famous General Scholium to Book Three, in which he attacked the vortex theory of Descartes, declared that the "most beautiful system of the sun, planets, and comets, could only proceed from the counsel and dominion of an intelligent and powerful Being," and discussed the nature of God, concluding: "And thus much concerning God; to discourse of whom from the appearance of things, does certainly belong to Natural Philosophy."[23] Newton then addressed himself to the problem of what gravitation is and how it might work, admitting that no assignment had been made of "the cause of this power" whose action explains the phenomena

of the heavens and the tides of the seas. This is
followed by the famous penultimate paragraph
which reads:

> But hitherto I have not been able to dis-
> cover the cause of those properties of
> gravity from phenomena, and I frame no
> hypotheses; for whatever is not deduced
> from the phenomena is to be called an hy-
> pothesis; and hypotheses, whether meta-
> physical or physical, whether of occult
> qualities or mechanical, have no place in
> experimental philosophy. . . . And to us it is
> enough that gravity does really exist, and
> act according to the laws which we have ex-
> plained, and abundantly serves to account
> for all the motions of the celestial bodies,
> and of our sea.

It was apparently the purpose of the General
Scholium to prevent any misunderstanding of
Newton's position such as had been made by
Bentley and Leibniz after reading the first edition of the *Principia* in which this General
Scholium did not appear. Yet the cautious word-
ing prevented the reader from gaining any insight into Newton's actual beliefs on this subject, as contained, for example, in a letter to
Boyle written on 28 February 1678/9, prior to
the publication of the *Principia* and not published until the mid-18th century. In this letter
Newton wrote out his speculations concerning

the "cause of gravity" and attempted to explain gravitational attraction by the operation of an all-pervading "aether" consisting of "parts differing from one another in subtility by indefinite degrees."[24] Some hint, but not more, of Newton's view was contained in the final paragraph of the above General Scholium, in which Newton wrote:

> And now we might add something concerning a certain most subtle spirit which pervades and lies hid in all gross bodies; by the force and action of which spirit the particles of bodies attract one another at near distances, and cohere, if contiguous; and electric bodies operate to greater distances as well repelling as attracting the neighboring corpuscles; and light is emitted, reflected, refracted, inflected, and heats bodies; and all sensation is excited, and the members of animal bodies move at the command of the will, namely, by the vibrations of this spirit, mutually propagated along the solid filaments of the nerves, from the outward organs of sense to the brain, and from the brain into the muscles. But these are things that cannot be explained in few words, nor are we furnished with that sufficiency of experiments which is required to an accurate determination and demonstration of the laws by

which this electric and elastic spirit op-
erates.

Thus the 18th-century reader who had become
convinced that the system of Newton's *Principia*
accounted for the workings of the universe and
then naturally enough wondered what the cause
of gravity might be, was tantalized by the final
statement that Newton might have elucidated
this topic but had decided not to do so. Hungry
for a further discussion of the nature of "this
electric and elastic spirit," or "aether," so intimately associated with the behaviour of material bodies and the communication of animal
sensation, such a reader would devour anything
else that Newton wrote on the subject.

Since Newton devoted a considerable portion
of the *Opticks* to this question, neatly avoided
in the *Principia,* we can understand at once why
the *Opticks* must have exerted so strong a fasci-
nation on men like John Locke and on all the
others who wanted to know the cause of gravity
and the fundamental principle of the universe.
Indeed, in the 1717 edition of the *Opticks*
Newton inserted an "Advertisement" (printed
below) explicitly declaring that he did "not take
Gravity for an Essential Property of Bodies,"
and noting that among the new Queries or Questions added to the new edition was "one Question concerning its Cause," Newton "chusing to
propose it by way of a Question" since he was

"not yet satisfied about it for want of experiments."

The first edition of the *Opticks* was published
in English in 1704 and contained only the first
16 queries. The Latin version, prepared at Newton's suggestion by Samuel Clarke, was issued
two years later in 1706; to it were added Queries 17-23 in which Newton discussed the nature
of the aether.[25] Here he compared the production of water waves by a stone to the vibrations
excited in the refracting or reflecting medium by
particles of light (Qu. 17), and proposed the
theory that the vibrations so excited "overtake
the Rays of Light, and by overtaking them successively, put them into the fits of easy Reflexion
and easy Transmission. . . ." He also suggested
that the aether is a "much subtiler Medium than
Air" and that its vibrations convey heat, both
radiant heat which may pass through a vacuum
and the heat communicated by hot bodies "to
contiguous cold ones" (Qu. 18); it is also "more
elastick and active" than air. This aether (Qu.
19, 20) is responsible for refraction and, because of its unequal density, also produces "Inflexions of the Rays of Light" or diffraction of
the sort recorded by Grimaldi. Newton indicated
(Qu. 21) his belief that the variations in density of the aether are the cause of gravitation;
and he pointed out that this aether must be
highly elastic to support the enormous speed of

light, but that the aether does not (Qu. 22) interfere with the motions of planets and comets;
and he compared the gravitational action of
the aether to the action of electrified and magnetic bodies in attracting, respectively, "leaf
Copper, or Leaf Gold, at . . . [a] distance . . .
from the electrick Body. . . . And . . . through
a Plate of Glass . . . to turn a magnetick
Needle. . . ." The last of these new Queries (Qu.
23) relates the vibrations of the aether to vision
and possibly to hearing.

One of the most interesting aspects of this
group of Queries is that it follows in outline the
letter written to Boyle some 20 years earlier,
and which had not yet been published. While
Newton was apparently willing to print his conjectures on the aether and gravitation in the
Queries of the *Opticks,* when he came to revise
the *Principia* for the second edition of 1713,
seven years after the appearance of the Latin
version of the *Opticks,* he was much more cautious. He refused, as we saw above, to discuss
the cause of gravitation, begging off with the
phrase *Hypotheses non fingo,*[26] and he actually
declared that all one needs to know is that gravity exists, that it follows the law of the inverse
square, and that it serves to describe the motions of the celestial bodies and the tides. Newton concluded the General Scholium with the
statement that he might discuss the aether but
refrained from doing so. While it can never be

proved, it is possible that Newton's procedure in this matter may indicate an appreciation of the fundamentally differing character of his two books—the *Principia* with its mathematical demonstrations and general avoidance of speculation, and the *Opticks* with its large speculative content.

To be sure, the speculations of the *Opticks* were not hypotheses, at least to the extent that they were framed in questions. Yet if we use Newton's own definition, that "whatever is not deduced from the phenomena is to be called an hypothesis," they are hypotheses indeed. The question form may have been adopted in order to allay criticism, but it does not hide the extent of Newton's belief. For every one of the Queries is phrased in the negative! Thus Newton does not ask in a truly interrogatory way (Qu. 1): "Do Bodies act upon Light at a distance . . .?" —as if he did not know the answer. Rather, he puts it: "Do not Bodies act upon Light at a distance . . . ?"—as if he knew the answer well— "Why, of course they do!" But if the addition of the question mark made it possible for Newton to free himself from the restrictions imposed by that "sufficiency of experiments which is required to an accurate determination and demonstration of the laws," we can only be grateful for the opportunity to see the mind of Newton at work, and to share his profound inspiration. As it was said in the 18th century,

any of the fine accounts of the Newtonian sys-
tem of celestial mechanics—such as those of
Pemberton, Maclaurin, and Voltaire—could af-
ford the reader a sufficient understanding of the
Principia which, therefore, need not be read in
Newton's original; but when it came to the
Opticks, no *abrégé* or *vulgarization* could take
the place of Newton's own words.

A glance at the analytical table of contents
will show the range of subjects covered by Newton in the *Opticks* and especially in the Queries.
For the latter not only took up questions of
light as such, and gravitation, but also chemis-
try, pneumatics, physiology, the circulation of
the blood, metabolism and digestion, animal
sensation, the Creation, the flood, the true nature of scientific inquiry, how to make experiments and how to draw the proper conclusions
from them by induction, the relation of cause
and effect, and natural philosophy in relation to
moral philosophy. Here, then, was a rich intel-
lectual feast for philosophers as well as scientists, for poets as well as experimenters, for
theologians as well as painters, and for all ama-
teurs of the products of the human imagination
at its highest degree of refinement. Not only,
therefore, did the *Opticks* come to enjoy a special place in 18th-century science, but also it had
an appeal wholly its own for the British poets
of the 18th century. Miss Marjorie Nicolson, in
the course of her investigations of science and

the literary imagination, arrived independently
at the same conclusion concerning the *Opticks*
and the 18th-century literary writers that I had
found to be true of the *Opticks* and the 18th-
century experimenters. "While reading widely
in eighteenth-century poetry for other purposes," she writes, "I found myself constantly
teased by dozens of references to Newton which
had nothing to do with the *Principia,* until I
became persuaded that, among the poets, the
Opticks was even more familiar than the more
famous work."[27]

One reason why the *Opticks* should enjoy a
greater vogue among experimenters than the
Principia is plain: the *Opticks* was, in great part,
an account of experiments performed by Newton and the conclusions drawn by him from the
experimental results; but in the *Principia* Newton described only two or three important experiments that he actually had made[28] and, for
the rest, merely cited data obtained by contemporaries and predecessors. We must remember,
furthermore, that throughout the history of science there have been two types of investigator:
the theoretician and experimenter, characterized
in modern times by such basic figures as Einstein
and Rutherford. It is rare that the two are combined within one individual as they were in
Newton. Newton the mathematical physicist of
the *Principia* did not in general appeal to those
who, by the application of the imagination and

the experimental art, were exploring new fields depending on empirical investigation for their future progress: such as plant physiology, chemical reactions, heat, and electricity. The mentor and guide of those who explored these new fields was Newton the heroic experimenter of the *Opticks* and the author of the Queries; and many of them, such as Franklin, who read and re-read the *Opticks,* did not even have the mathematical training to attempt to read the *Principia,* had they wanted to do so.

From the point of view of the 18th century, as indeed from that of the 19th, Newton's *Principia* with its law of universal gravitation had apparently, but for certain minor revisions and emendations, settled the problem of heavenly motions once and for all. Certain situations were not soluble; for example, the famous three-body problem does not admit of a general analytic solution. Yet the law of universal gravitation was unquestionably true, something to be believed even when all else failed. This faith is mirrored for us in a letter which the naturalist Thomas H. Huxley wrote, in 1860, after the death of his son, "I know what I mean when I say I believe in the law of the inverse squares, and I will not rest my life and my hopes upon weaker convictions."[29]

But while the *Principia* seemed the terminal point of an ancient line of inquiry, the *Opticks,* with its newly discovered phenomena concern-

ing colors and diffraction, clearly marked the
beginning of a new direction in physical inquiry.
Whereas Newton could end the *Principia* with a
General Scholium and supplement it with *The
System of the World,* he closed the *Opticks* on a
note of uncertainty, on a set of Queries—of
which some, but by no means all, might be resolved by the work of future generations.

The point was made earlier that Newton's
optical researches are related to our present
views, not by a conscious return of 20th-century
physicists to the Newtonian ideas, but rather
through a chain that extends from one end of
the 19th century to the other. The 19th century
produced a succession of brilliant research, of
which the crowning achievement was Clerk
Maxwell's electromagnetic theory of light. In
the end, it was the partial failure of this theory
that led to the quantum theory of Planck and
the photon theory of Einstein. The electro-
magnetic theory derives in part from the exten-
sion by Clerk Maxwell of Faraday's notions
concerning the propagation of electric and magnetic effects through space,[30] and it brought to
a superlative conclusion the wave theory of
light, revived by Thomas Young in the opening
years of the century.

Young's initial contributions were made as a
result of studying a class of optical phenomena
which we call today effects of interference and
diffraction, but which Newton called the "in-

flection" of light. The first major account of
such phenomena was that published by F. M.
Grimaldi in his *Physico-Mathesis de Lumine
Coloribus et Iride* (Bologna 1665). Further
studies of diffraction were made by Boyle and
Hooke as well as Newton, but the most signifi-
cant and quantitative results were those obtained by Newton while studying the interference rings produced when the curved surface
of a plano-convex lens was pressed against a
flat optical surface. Newton's magnificent experiments, described in the text in Book Two,
provided conclusive evidence that some kind of
periodicity is associated with the several colors
into which he divided visible light. Such periodicity can in no way be accounted for by the
mechanical action of corpuscles moving in right
lines and Newton was, therefore, forced into
the position of having to postulate some kind of
waves accompanying the corpuscles; these were
the famous aether waves. Newton had to account for the successive refraction and reflection that he supposed must occur at the glass-air
interfaces between a convex and a flat optical
surface when the interference rings are produced. He suggested that the alternate "fits of
easy reflection" and "fits of easy refraction"
arise from the action of the aether waves which
overtake the particles of light and put them into
one or the other state.

Newton's measurements of the separation of
these rings and his computations of the thickness of the thin film of air between the two glass
surfaces were of the highest order of accuracy.
After Young had explained the production of
"Newton's rings" by the application of his new
principle of interference to the wave theory of
light, he used Newton's data to compute the
wave-length of different colors in the visible
spectrum and also the wave-numbers (in "Number of Undulations in an Inch") ; Young's computations,[31] based on Newton's measurements,
yielded a wave-length for the "Extreme red" of
0.000, 026, 6 inches, or a wave-number of 37640
"undulations in an inch" and a frequency of
436 x 10^6 "undulations for a second"; and for
the "extreme violet" the same quantities had
values 0.000, 016, 7 inches, 59750 per inch, and
735 x 10^6 per sec, respectively, in close agreement with present-day accepted values.[32]

Young was indebted to Newton for more
than the data for computing wave-lengths, wave-
numbers, and frequencies; the whole wave theory of light was developed by him from the suggestions in Newton's *Opticks,* with several important additions, chiefly (1) considering the
waves in the aether to be transverse, (2) supple-
menting the wave theory by the principle of interference. It was from Young's work that the
19th-century developments leading to the electromagnetic theory may be said to have begun;

and since Young's work was inspired by Newton's, we have an historical chain leading from Newton to Young, and from Young to Fresnel and Arago, and from them to Clerk Maxwell and eventually to Planck and Einstein.

Young was extremely explicit about his debt to Newton. Thus, in the first of the three foundational papers in the *Philosophical Transactions,* Young stated:

> The optical observations of Newton are yet unrivalled; and, excepting some casual inaccuracies, they only rise in our estimation as we compare them with later attempts to improve on them. A further consideration of the colours of thin plates, as they are described in the second book of Newton's Optics, has converted the prepossession which I before entertained for the undulatory system of light, into a very strong conviction of its truth and sufficiency; a conviction which has been since most strikingly confirmed by an analysis of the colours of striated substances. . . .
>
> A more extensive examination of Newton's various writings has shown me that he was in reality the first that suggested such a theory as I shall endeavour to maintain. . . .

Young even pointed out that the wave theory of light should be given a hearty welcome because

it originated with Newton who was universally
venerated.

> Those who are attached, as they may be
> with the greatest justice, to every doctrine
> which is stamped with the Newtonian ap-
> probation, will probably be disposed to
> bestow on these considerations so much the
> more of their attention, as they appear to
> coincide more nearly with Newton's own
> opinions. For this reason, after having
> stated each particular position of my the-
> ory, I shall collect, from Newton's various
> writings, such passages as seem to me the
> most favourable to its admission. . . .

In conformity with this plan, almost half of the
article is made up of quotations from Newton.
In his "Reply to the Edinburgh Reviewers,"[33]
Young described the history of his ideas concerning light, once again expressing the degree
to which they derived from Newton's writings.
Even the principle of interference, he insists,
was discovered by him in May 1801 "by reflecting on the beautiful experiments of Newton . . . ," and although "there was nothing that
could have led to it in any author with whom
I am acquainted," Young noted how Newton
had used a similar conception in explaining "the
combinations of tides in the Port of Batsha."[34]
Although there is, therefore, no question of the
influence of Newton on Young, one suspects that

Young, like the Player Queen in Hamlet, doth
protest too much. The *Edinburgh Review,* in its
condemnation of Young, was of course grossly
unfair when it used against him the phrase that
a person may, "with the greatest justice, be
attached to every doctrine which is stamped with
the Newtonian approbation," but without quotation marks (as if this were a position *contra*-
Young on the part of a *pro*-Newton reviewer
rather than a statement by Young himself);
but Young should not really have been as aston-
ished as he would have us believe he was; after
all, in the minds of everyone at that time, Newton stood for the corpuscular theory of light and
against the theory of light waves.

This raises for us the very real question of
whether we can, with Young and Preston, strip
Newton's theory of the corpuscles and leave
only the waves as the essential components. And
this question rapidly leads us into another,
which is, why did Newton build his theory on
corpuscles, or, why did he reject the wave theory
that others in the 19th century in vain tried to
attribute to him?

I shall not attempt to trace here the complete
history of Newton's ideas concerning the theory
of light, nor even to collate all the evidence relating to this subject that may be found in his
various published works and letters, but shall
rather limit myself to the information provided
in the *Opticks* itself. Foremost among the rea-

sons why Newton insisted upon the corpuscular-
ity of light was the general atomism of the age;
indeed, the very hall-mark of the "New Science" in the 17th century, among such men as
Bacon, Galileo, Boyle, and others, was a belief
in atomism, in what Boyle called the "corpuscular philosophy." Whereas the scholastic doctrine had placed light and the phenomena of
colors in the category of "forms and qualities,"
men such as Newton opposed to this traditional
view an explanation of the phenomena of nature
in terms of the mechanical action of atoms, or of
matter and motion. Summing up the many reasons for a general belief in atoms in the final
Query, Newton wrote:

> All these things being consider'd, it
> seems probable to me, that God in the
> Beginning form'd Matter in solid, massy,
> hard, impenetrable, moveable Particles, of
> such Sizes and Figures, and with such
> other Properties, and in such Proportion
> to Space, as most conduced to the End for
> which he form'd them. . . .
> Now by the help of these Principles, all
> material Things seem to have been com-
> posed of the hard and solid Particles
> above-mention'd, variously associated in
> the first Creation by the Counsel of an in-
> telligent Agent. For it became him who
> created them to set them in order. And if

he did so, it's unphilosophical to seek for
any other Origin of the World. . . .

Then, too, it was well known that waves of
whatever sort would spread out in all directions
in any homogeneous medium, rather than travel
in straight lines as light is observed to do when
it produces a sharply defined shadow. Thus, says
Newton (Qu. 29) : "Are not the Rays of Light
very small Bodies emitted from shining Substances? For such Bodies will pass through uniform Mediums in right Lines without bending
into the Shadow, which is the Nature of the
Rays of Light." Furthermore, that material
bodies moving in a straight line oblique to a
surface will be reflected so as to obey the law
of reflection, i.e. that the angle of incidence
equals the angle of refraction, had been well
known since classical antiquity. Refraction might
easily be explained on the basis of the corpuscular theory since the attraction exerted by the
particles of glass, say, on the corpuscles of light
incident upon the glass from air would produce
an increase in the vertical component of the
velocity of the particles and, therefore, would
result in a bending toward the normal which
is always observed to be the case.[35]

Of course, a rival theory to Newton's, the
wave theory of Christiaan Huygens, had offered
a geometrical construction for reflection and re-

fraction in wholly different terms. And this theory led to an exactly opposite conclusion to that of Newton's theory in respect to the relative speeds of light in air and in glass or water. Whereas Newton's corpuscular theory demands a speed of light greater in glass or water than in air, the theory of Huygens requires a speed of light in air that must be greater than the speed of light in water or in glass. Unfortunately, the possibility of putting these opposing conclusions to the test of experiment did not occur until well into the 19th century, when the labors of Young and Fresnel had already established the wave theory of light, so that this test, favoring the conclusions of the wave theory rather than the corpuscular, was but an additional argument, rather than the primary one, for the wave theory of light.

Newton's rejection of Huygens' theory was based, in part, on his own cherished belief in atomicity and also on the fact that Huygens' theory was geometrical rather than mechanical, and contradicted physical principles. Although Huygens had provided a brilliant method for constructing the wave front in the case of refraction or reflection, the waves he postulated were without the primary characteristic of a physical wave motion, i.e., periodicity. In fact, Huygens' denial of periodicity in his postulated light waves was an attempt to account for the possibility of a number of waves crossing each

other without in any way interfering one with
another. But without the property of periodicity,
such waves could not account for color, nor
any of the other periodic properties of light
which Newton had observed in the various types
of interference and diffraction phenomena he
had studied so carefully in the *Opticks*. Nor
could Huygens, without the principle of de-
structive interference invented by Young a little
more than a hundred years later, adequately account for the simplest of all optical phenomena,
rectilinear propagation.

Finally, the most brilliant of all the portions
of Huygens' *Treatise on Light*[36] provided Newton
with an argument against Huygens' theory.
For, by extending the geometric construction of
wave fronts from isotropic to anisotropic media,
Huygens had been able to account for the phe-
nomenon of double refraction in calcite, or Ice-
land spar, by two different wave forms. Newton
(Qu. 28) considered this to be an important
weapon against Huygens' "Hypothesis." Newton grasped the salient aspect of Huygens' investigation, which was that "the Rays of Light
have different Properties in their different
Sides," and he quoted from the original French
of Huygens to prove how baffling this phe-
nomenon was to the author of the wave theory
himself; plainly, "Pressions . . . propagated
. . . through an uniform Medium, must be on
all sides alike." It never apparently occurred to

Huygens, who thought in terms of a geometric scheme, nor to Newton, that the undulations might be perpendicular to the direction of propagation. When, eventually, it was suggested by Young and Fresnel that light waves must be transverse rather than longitudinal, then for the first time was it possible to explain the polarization of light, or the way in which light—to use Newton's phrase—has "sides." The study of the interference of polarized beams of light provided in the 19th century one of the chief arguments for the advocates of the wave theory. But in Newton's day and for a hundred years thereafter, the only way to account for the "sides" of light was to suppose that the corpuscles were not perfectly spherical and would present, therefore, different sides depending on their orientation to the axis of motion.

It is one of the ironies of history that the *Opticks,* based on Newton's secure belief that light rays consist of streams of corpuscles, should have provided, a century later, as great a contribution to the wave theory as the *Traité de la lumière* of his rival Huygens. And it is just this feature which makes the *Opticks* such an attractive work for anyone to read who is interested in that stage of creation when the greatest and clearest of minds is baffled by the observed data and cannot reduce them to a simple, clear, straightforward conceptual scheme. As we watch Newton wrestle with the problems of the nature of light, we get in the following pages some measure of the extraordinary difficulty of scientific research applied to the most fundamental problems. And if Newton's elegant conceptions do not always permit us to follow his argument in every detail, we can at any rate be grateful for the opportunity to observe in action one of the greatest minds that science has ever produced, so beautifully described by Wordsworth's "Newton with his prism and silent face," his "mind forever voyaging through strange seas of thought alone."[37]

I. BERNARD COHEN

[1]From the version edited by T. Oliver Harding: *Handbook of Natural Philosophy: Optics,* sixth thousand, London, James Walton, 1869, p. 164.

[2]Arthur Schuster: *An Introduction to the Theory of Optics,* second edition, revised, London, Edward Arnold, 1909, p. 86.

[3]These three memoirs, together with other writings on the subject by Young, may be found in George Peacock, editor, *Miscellaneous Works of the Late Thomas Young,* London, John Murray, 1855. "On the Theory of Light and Colours" was a Bakerian Lecture read at the Royal Society on 12 Nov. 1801 and was published in the *Phil. Trans.* for 1802, pp. 12 ff. (reprinted in *Misc. Works,* vol. 1, pp. 140 ff.), "An Account of some

Cases of the Production of Colours not hitherto described," was read at the Royal Society on 1 July 1802 and was published in the *Phil. Trans.* for 1802, pp. 387 ff. (reprinted in *Misc. Works,* vol. 1, pp. 170 ff.), and "Experiments and Calculations relative to Physical Optics" was a Bakerian Lecture read at the Royal Society on 24 Nov. 1803 and was printed in the *Phil. Trans.* for 1804, pp. 1 ff. (reprinted in *Misc. Works,* vol. 1, pp. 179 ff.).

[4]By "unassailable," I do not of course mean that the wave theory, as an exclusive explanation of optical phenomena, has remained unassailable; but rather that the application of the principle of interference to diffraction phenomena has remained the major basis for our belief in waves, whether these be matter waves, probability waves, electro-magnetic waves, or old-fashioned aether waves. Thus, when Louis de Broglie predicted the existence of matter waves having a wave-

length $\lambda = \dfrac{h}{mv}$, evidence for their existence in the case

of electrons was provided by the diffraction pattern produced by a crystalline metal on a beam of thermal electrons—in accordance with the application of the principle of interference in much the same manner as it had been applied to visible light in Young's pin-hole experiment and to X-rays in Laue's crystal experiments.

[5]The attack on Young's three memoirs was published in the *Edinburgh Review,* nos. II and IX. The authorship of the attack is discussed by Peacock in the note added to the reprint of Young's reply (*Misc. Works,* vol. 1, pp. 192 ff.).

[6]Cf. Florian Cajori, *A History of the Conceptions of Limits and Fluxions in Great Britain from Newton to*

Woodhouse, Chicago, Open Court Pub. Co., 1919.

[7]Thomas Preston, *Theory of Light,* fifth edition, edited by Alfred W. Porter, London, Macmillan, 1928, p. 21; the preface to the first edition is dated July 1890.

[8]This research has been sponsored by a generous grant in aid from the American Philosophical Society. A preliminary report has been published in *The American Philosophical Society Year Book 1949,* Philadelphia, 1950, pp. 240-243. It is hoped that a long memoir, embodying the results of research on the nature of physical theory in the 18th century and its effect on that of the 19th, stressing the role of Newton's *Opticks* and Franklin's unitary theory of electrical action, will be published by the American Philosophical Society late in 1952.

[9]"Vegetable Staticks," vol. 1 of *Statical Essays,* ed. 2, London, W. Innys et al., 1731. An important study of Hales, clearly indicating the indebtedness of Hales to the Queries in the *Opticks,* is Henry Guerlac: "The Continental Reputation of Stephen Hales," *Archives Internationales d'Histoire des Sciences,* 1951, No. 15, pp. 393-404.

[10]Hales refers to Motte's abridgment of the *Phil. Trans.,* vol. 2, p. 1.

[11]This may also be found in Qu. 31 of the *Opticks;* see pp. 380-381 below.

[12]The *Principia* is available in Florian Cajori's version of Andrew Motte's English trans. of 1729, Berkeley, Univ. of Calif. Press, 1934.

[13]Craigie's letter to Bentley, dated 24 June 1691, may be found in Sir David Brewster, *Memoirs of the Life, Writings, and Discoveries of Sir Isaac Newton,* Edinburgh, Thomas Constable, 1855, vol. 1, p. 465.

[14]*Ibid,* p. 464.

[15]Cf. preface to Jean-Théophile Desaguliers, *Course of Experimental Philosophy,* ed. 3, vol. 1, London, A. Millar, 1763, p. viii.

[16]*Ibid.*

[17]A Latin edition of the *Opticks* was later prepared at Newton's request by Samuel Clarke and published in London in 1706, two years after the English edition.

[18]J. Edleston, ed., *Correspondence of Sir Isaac Newton and Professor Cotes* . . . , London, John W. Parker, 1850, p. 153.

[19]This letter, in the original Latin, may be found in Newton's *Opera,* ed. by Samuel Horsley, vol. 4, London, John Nichols, 1782, pp. 314 ff. The English translation is quoted from Cajori's notes to his edition of the *Principia,* ed. cit., p. 673. Horsley printed the entire letter; when originally published in the *Phil. Trans.,* this part of the letter had been omitted.

[20]The Newton-Bentley correspondence may be found in Newton's *Opera,* ed. by Horsley, vol. 4, pp. 427 ff.

[21]*Correspondence of Sir Isaac Newton and Professor Cotes,* p. 153. [Cotes told Newton that the Leibniz letter appeared in "the 18th Number of the second Volume" of the *Memoirs of Literature.* In the second edition of the *Memoirs of Literature,* "revised and corrected," Leibniz's letter appears in vol. 4 (London, R. Knaplock, 1722), art. LXXV, pp. 452 ff.]

[22]*Ibid.*

[23]All quotations from the *Principia* are taken from Cajori's edition (note 12).

[24]Newton's letter to Boyle, dated 28 Feb. 1678/9 was first printed in Boyle's *Works,* edited by Thomas Birch, vol. 1, London, A. Millar, 1744, Life, pp. 70 ff.

The aether, in respect of its rarity and density, not only was supposed to produce gravitation but also the diffraction phenomena described by Grimaldi. This letter is reprinted in Sir David Brewster, *Memoirs,* ed. cit. (note 13).

[25]These Queries, 17-23, next appeared in the second English edition, first issued in 1717, together with new Queries 24-31.

[26]An important discussion of Newton's use of hypotheses, and the meaning of the phrase *Hypotheses non fingo,* by Cajori may be found in his edition of the *Principia,* p. 671.

[27]Marjorie Hope Nicolson, *Newton demands the Muse: Newton's "Opticks" and the Eighteenth Century Poets,* Princeton, Princeton Univ. Press, 1946, p. vii.

[28]Three major experiments performed by Newton and described at length in the *Principia* are (1) the production of interference fringes by the diffraction of light produced by the "edges of gold, silver, and brass coins, or of knives, or broken pieces of stone or glass," as "lately discovered by Grimaldi," Scholium following prop. XCVI, th. L, bk. 1 (pp. 229 ff., Cajori's ed.); this Scholium occurs in a group of theorems in which is shown "the analogy there is between the propagation of the rays of light and the motion of bodies," but "not at all considering the nature of the rays of light, or inquiring whether they are bodies or not"; (2) a study of the "resistance of mediums by pendulums oscillating therein," General Scholium following prop. XXXI, th. XXV, bk. II (pp. 316 ff.); (3) an investigation of "the resistances of fluids from experiments," consisting of measurements of the time of descent of spheres made of (or filled with) various substances, when allowed to

fall through water and air, Scholium following prop. XL, prob. IX, bk. 2 (pp. 355 ff.).

[29]The letter, written to Charles Kingsley, may be found in *Life and Letters of Thomas Henry Huxley*, edited by Leonard Huxley, vol. 1, London, Macmillan, 1900, p. 218.

[30]Faraday used to quote Newton's letter to Bentley as inspiration for his own concern over the role of the medium in the action of one body upon another at a distance from it. Cf. John Tyndall: *Faraday as a Discoverer*, new ed., London, Longmans, Green, 1870, p. 81.

[31]"On the Theory of Light and Colours" (see footnote 3 above) in *Misc. Works*, vol. 1, p. 161.

[32]Newton never computed wave-lengths, but on many occasions noted that the vibrations of the "aether" corresponding to the several colors might be likened to the vibrations of air which, according to their several "bignesses, makes several Tones in Sound." One may conclude from the analogy of sound to light (Qu. 28) that Newton's aether waves would be longitudinal pulses; yet Newton also referred to the mode of production of water waves by a stone thrown into a stagnant water (Qu. 17) as an analogy to the "aether" waves arising from the motion of light-particles, and in this case the disturbance would be transverse.

[33]Reprinted in *Misc. Works*, vol. 1, pp. 192 ff.

[34]This example is discussed more fully in I. B. Cohen, "The First Explanation of Interference," *Am. J. Physics* (1940), vol. 8, pp. 99 ff.

[35]In one of his earliest publications in optics, Newton had already indicated that rectilinear propagation was contrary to the suppositions of a wave theory: "For, to

me, the Fundamental Supposition it self seems impossible; namely, That the *Waves* or Vibrations of any Fluid, can, like the Rays of Light, be propagated in *Streight* lines, without a Continual and very extravagant spreading and bending every way into the Quiescent Medium, where they are terminated by it. I mistake, if there be not both Experiment and Demonstration to the contrary." (*Phil. Trans.*, no. 88, p. 5089). In the *Principia*, Prop. XLII, th. XXXIII, bk. II, there is a demonstration that "All motion propagated through a fluid diverges from a rectilinear progress into the unmoved spaces." The scholium to Prop. L, Problem XII, bk. II, deals with the determination of the wavenumber of a sound-vibration, but begins with a declaration that the rectilinear propagation of light proves that light cannot consist of waves alone.

In Query 28, Newton indicated by examples that "Waves, Pulsations or Vibrations of the Air, wherein Sound consists, bend manifestly, though not so much as the Waves of Water." Does not the evidence of the bending of light in diffraction experiments, as described in the beginning of Book Three, indicate the wave nature of light? Although Newton referred to such experiments (p. 388) as examples of "how the Rays of Light are bent in their passage by Bodies," they were interpreted as illustrating not wave motion or "pression" so much as the way (Qu. 1) "Bodies act upon Light at a distance, and by their action bend its Rays." In Query 28, the difference between the bending of light and of waves is discussed; we find, "The Rays which pass very near to the edges of any Body, are bent a little by the action of the Body, as we shew'd above; but this bending is not towards but from the Shadow, and is perform'd only in the passage of the Ray by the Body, and at a very small distance from 'it.''

[36] Huygens' work is available in an English version by S. P. Thompson as *Treatise on Light,* London, Macmillan, 1912; reprinted by the University of Chicago Press.

[37] In addition to the works mentioned in the footnotes, a few others may be cited for those who may wish to pursue some aspect of this subject further. An excellent bibliographical study of the major publications of Newton and important commentaries may be found in *A Descriptive Catalogue of the Grace K. Babson Collection of the Works of Isaac Newton and the Material relating to him in the Babson Institute Library,* New York, Herbert Reichner, 1950, which supplements but does not entirely supersede George K. Gray: *A Bibliography of the Works of Sir Isaac Newton, together with a List of Books illustrating his Works,* second ed., Cambridge, Bowes and Bowes, 1907. A number of important and stimulating essays may be found in The Royal Society: *Newton Tercentenary Celebrations, 15-19 July 1946,* Cambridge, Cambridge Univ. Press, 1947, especially those by E. N. da C. Andrade, Lord Keynes, and S. I. Vavilov. George F. Shirras is writing a new biography which will have, among other merits, the benefit of study of many unpublished manuscript documents, a considerable portion of which were collected by his teacher, the late Lord Keynes. Alexandre Koyré is preparing a series of Newtonian Studies to match his *Etudes Galiléennes;* he has provided an earnest of this great work in "The Significance of the Newtonian Synthesis," *Archives Internationales d'Histoire des Sciences,* 1950, vol. 29, pp. 291-311. An admirable exposition of Newton's early work on color

and the prismatic spectrum is Michael Roberts and
E. R. Thomas, *Newton and the Origin of Colours,*
London, G. Bell & Sons, 1934.

FOREWORD

FORTUNATE Newton, happy childhood of science! He who has time and tranquillity can by reading this book live again the wonderful events which the great Newton experienced in his young days. Nature to him was an open book, whose letters he could read without effort. The conceptions which he used to reduce the material of experience to order seemed to flow spontaneously from experience itself, from the beautiful experiments which he ranged in order like playthings and describes with an affectionate wealth of detail. In one person he combined the experimenter, the theorist, the mechanic and, not least, the artist in exposition. He stands before us strong, certain, and alone: his joy in creation and his minute precision are evident in every word and in every figure.

Reflexion, refraction, the formation of images by lenses, the mode of operation of the eye, the spectral decomposition and the recomposition of the different kinds of light, the invention of the reflecting telescope, the first foundations of colour theory, the elementary theory of the rainbow pass by us in procession, and finally come his observations of the colours of thin films as the origin of the next great theoretical advance, which had to await, over a hundred years, the coming of Thomas Young.

Newton's age has long since been passed through the sieve of oblivion, the doubtful striving and suffering of his generation has vanished from our ken; the works of some few great thinkers and artists have remained, to delight and ennoble us and those who come after us. Newton's discoveries have passed into the stock of accepted knowledge: this new edition of his work on optics is nevertheless to be welcomed with warmest thanks, because it alone can afford us the enjoyment of a look at the personal activity of this unique man.

ALBERT EINSTEIN

INTRODUCTION

A HUNDRED years ago, the world thought it-self qualified to pass a final judgment on Newton's work in theoretical physics. His law of gravitation, regarded then and now as the greatest of all scientific discoveries, was held to be ultimate and unassailable, the typical law according to which all other laws must be fashioned; the whole of nature was eventually to be explained in terms of attractions and repulsions between particles moving in a vacuum. With regard to Newton's optical work, however, our grandfathers spoke with less enthusiasm; the brilliance of much of it was undisputed, but it was generally believed that Newton had compromised himself by favouring the corpuscular theory of light, and that the corpuscular theory was hopelessly wrong; and so, while portions of the *Principia* were still regularly studied by young men in the latter half of the nineteenth century, the *Optics* was treated rather as a book of interest only to the scientific historian.

How times have changed! In the middle of the nineteenth century, Faraday and Maxwell replaced Coulomb's law of force between electrified particles (which resembled the Newtonian law of gravitation) by a propagation of influences between contiguous particles of a universal medium; and since then the fortunes of action-at-a-distance have steadily declined. In 1915 Newton's law itself succumbed, and was superseded for theoretical (though not for practical) purposes by Einstein's theory of general relativity. Even Euclidean geometry has been objected to on the ground that it savours too much of *Fernwirkung*.

The development of optical theory during recent years has seemed to the older school of physicists even more astonishing. Corpuscular ideas of light, after having been dead for a century were resuscitated in 1905, when Einstein explained the photoelectric effect by aid of Planck's quantum principle, and so was led to postulate the existence of "quanta" of light. His conception has been fully confirmed by experiments and particularly by the "Compton effect," which shows that when a light-quant collides with an electron, the ordinary dynamical laws of impact are obeyed. At the same time, the older experiments by which light was shown to consist of waves, have not been invalidated; and thus we have been forced to recognise that the wave-hypothesis and the corpuscular hypothesis are *both* true.

Before 1927 this last assertion would have been regarded as a contradiction in terms; it presented in fact, a paradox like that which had appeared a generation earlier when Michelson and Morley proved that the motion of their laboratory relative to the aether was null at all times, although its motion as ordinarily understood was continually varying. The Michelson-Morley paradox was solved by making a closer examination of the fundamental notions of space and time, leading eventually to the theory of relativity; and similarly the paradox of the co-existence of the undulatory and corpuscular theories of light has been at any rate partially explained by making a closer examination of fundamental principles, leading to Heisenberg's law of the absolute limit of accuracy of measurements and to the realisation that a dual aspect, as waves and corpuscles, is to be expected not only in the case of light, but also in the case of electrons (which were formerly regarded only as corpuscles) and even in the case of atoms also.

All this has had its effect in modifying opinion regarding Newton's Optics. The curious blending of corpuscular-theory with wave-theory which is suggested in some parts of his work, and which was a stumbling-block to the physicists of the nineteenth century, has been found to present considerable analogies with the modern views; and indeed recently a distinguished physicist after describing some excellent work of his own, quoted at length the well-known passage forming the latter part of Proposition *XII* of the second book with the comment "After being regarded for generations as an artificial attempt to save a dying theory, we have proved this guess of Newton's to be a supreme example of the intuition of genius." So the volume which is here reprinted, after being esteemed for three generations chiefly as a historical landmark displaying a marvellous combination of theoretical and experimental skill, is now once more being read for its living scientific interest. In any case, it will always compel attention because its subject was the first, and also the last, in which Newton made important discoveries, and because it was the only considerable scientific work which he himself prepared and corrected for the press and enriched with new matter in successive editions. Let us now trace, so far as we can, its development in the mind of its author.

Newton's acquaintance with optics seems to have begun when he was an undergraduate at Cambridge by the reading of Kepler's *Dioptrice*, which had been published in 1611. The true law of refraction was then unknown, but it was assumed that the angle of refraction is proportional to the angle of incidence, which is equivalent to the true law when the incidence is nearly perpendicular; and by aid of this assumption Kepler had given a sound explanation of the performance of lenses and refracting telescopes. In the same year 1611 had appeared the *De radiis visus et lucis* of Antonio de Dominis, Archbishop of Spalato, which gave the first approximately correct explanation of the rainbow, and may have originated Newton's marked interest in atmospheric optics.

Less valuable from the point of view of scientific discovery, but perhaps more stimulating, were the works of Descartes, of which the *Dioptrique* and the *Météores* had been published at Leyden in 1638. In the last half of the seventeenth century indeed, the theories of Descartes were dominant. According to the Cartesian philosophy, all space—even at the remotest distances beyond the stars—is a plenum, so that a particle can move only by taking the place of other particles which themselves are displaced. Light was imagined to be essentially a pressure transmitted through this dense mass of particles; vision might therefore be compared to the perception of the presence of objects which a blind man obtains by the use of his stick, the transmission of pressure along the stick from the object to the hand being analogous to the transmission of pressure through the plenum of space from a luminous object to the eye.

Descartes supposed "the diversities of colour and light" to be due to the different ways in which the matter moves: the various colours were connected with different rotatory velocities of the particles, those which rotate most rapidly giving the sensation of red, the slower ones of yellow, and the slowest of green and blue.

Beside the books which have been mentioned, Newton had at Cambridge the oral teaching of Barrow, Lucasian Professor from 1664, who delivered lectures on Optics which were afterwards (in 1669) published. They are in many ways admirable, but at the end of the twelfth lecture Barrow explains his ideas on colour, which are very crude: *e.g.* red is that which sends out light more concentrated than usual, but broken up by dark interstices: yellow consists mostly of white with some red interspersed, and so on.

Three other notable works appeared at about the time when Newton was beginning his independent researches, namely the *Optica Promota* of James

Gregory (1663) which described his reflecting telescope; the *Physico-Mathesis* of the Jesuit Father
Grimaldi (1665) which contained his discovery of
diffraction; and the *Micrographia* of Robert Hooke
(1665-7) in which he described the colours of thin
plates, and explained them by introducing the idea
of interference, on which long afterwards the complete theory was based. Regarding colour and the
nature of light, he advanced the view that light is a
small and rapid vibratory motion of the medium, and
that colour depends on the form of the light-pulse:
"blue is an impression on the retina of an oblique
and confused pulse of light, whose weakest part precedes and whose strongest follows"; while in red, the
strongest part precedes and the weakest follows.

Newton's own activity as an original worker
in optics seems to have begun in 1663, when he
began grinding lenses and interesting himself in
the construction and performance of telescopes.
Feeling the desirability of reducing or removing
the chromatic aberration, at the beginning of
1666 (being then aged twenty-three) he bought a
glass prism "to try therewith the phenomena of
colours." For this purpose "having darkened my
chamber, and made a small hole in my window-shuts,
to let in a convenient quantity of the Sun's light, I
placed my Prisme at his entrance, that it might be
thereby refracted to the opposite wall." Observing
that the length of the coloured spectrum was many
times greater than its breadth, he was led after more
experiments to the view that ordinary white light

is really a mixture of rays of every variety of colour, and that the elongation of the spectrum is due to the differences in the refractive power of the glass for these different rays. "To the same degree of Refrangibility ever belongs the same colour, and to the same colour ever belongs the same degree of Refrangibility."

This discovery was the subject of his first scientific paper, published in the Royal Society's *Transactions* in 1672. Its appearance gave rise to an acute controversy. Hooke in particular assailed it vehemently; and the unpleasant consequences which followed its announcement had much to do with the reluctance which Newton ever afterwards showed to make known his results to the world. As his disciple Maclaurin said, "The warm opposition his admirable discoveries met with, in his youth, deprived the world of a full account of them for many years, till there appeared a greater disposition amongst the learned to receive them, and induced him to retain other important inventions by him, from an apprehension of the disputes in which a publication might involve him." At the time indeed he seriously contemplated abandoning research altogether. "I intend," he wrote to Oldenburg, "to be no farther solicitous about matters of Philosophy: and therefore I hope you will not take it ill, if you find me never doing anything more in that kind."

In the course of the discussion Newton took occasion to explain more fully his views on the nature of light. Hooke charged him with holding the doc-

trine that light is a material substance. Now Newton
had, as a matter of fact, a great dislike of the more
speculative kind of hypotheses; his aim was to create
a theory based directly on observation and free from
all imaginings as to the hidden mechanism of
things. "He used," says Maclaurin, "to call his philosophy *experimental philosophy*, intimating, by the
name, the essential difference there is betwixt it and
those systems that are the product of genius and
invention only." "Hypotheses," said Newton himself,
"are not to be regarded in experimental Philosophy."
Accordingly, in reply to Hooke's criticism, he pro-
tested that his views on colour were in no way bound
up with any particular conception of the ultimate
nature of optical processes. However, in order to
relate them to Hookes' hypotheses, and having made
sure that colour is an inherent characteristic of light,
he inferred that it must be associated with some
definite quality of the corpuscles or aether-vibrations.
The corpuscles corresponding to different colours
would, he remarked, like sonorous bodies of different
pitch, excite vibrations of different types in the
aether: and "if by any means those (aether-vibrations) of unequal bignesses be separated from one
another, the largest beget a sensation of a *Red* colour,
the least or shortest of a deep *Violet*, and the intermediate ones, of intermediate colours."

This sentence is the first enunciation of the principle that homogeneous light is essentially *periodic*
in its nature, and that differences of period corre-
spond to differences of colour. The analogy with

Sound is obvious: "much after the manner that the Vibrations of the Air, according to their several bignesses, excite sensations of several Sounds," he says: and it may be remarked in passing that Newton's theory of periodic vibrations in an elastic medium, which he developed in connexion with the explanation of Sound, would alone entitled him to a place among those who have exercised the greatest influence on the theory of light, even if he had made no direct contribution to the latter subject.

It was extraordinarily difficult for even the most brilliant of Newton's contemporaries to accept his doctrine that light existed in an infinite number of different independent colours, incapable of being changed into each other, and characterised by a definite refrangibility. He seemed to be refuted by experiments in which a mixture of paints of two colours produced paint of a third colour, and by other experiments belonging to the subjective physiological theory of colour-vision, in which colours really are compounded out of primaries. Even Huygens in 1673 said that "an hypothesis which should explain the colours yellow and blue, would be sufficient for all the rest."

The discovery that ordinary white light is a heterogeneous mixture of differently refrangible rays led Newton to realise more fully the difficulty of getting rid of chromatic aberration in lenses. Believing that all transparent bodies have the same dispersive power, he missed the discovery of achromatism; and concluded that "the improvement of telescopes of given lengths by refractions is desperate"; which led him to turn to reflection rather than refraction as the basic principle of optical instruments. In 1668 he invented the type of reflecting telescope which bears his name, in which a small plane mirror inclined at $45°$ to the axis of the instrument intercepts the rays reflected by the large mirror a little before they arrive at the focus, and so throws them to the side of the tube, where they enter the eyepiece: an example was sent to the Royal Society in 1671. In the description, *An Accompt of a new Catadioptrical Telescope*, it is stated that "the objects are magnified about 38 times," whereas "an ordinary telescope of about two feet long only magnifies 13 or 14 times."

At the end of 1675 Newton sent to the Royal Society a paper in which, departing somewhat from his usual attitude of antagonism to hypotheses, he indicated that to which at the time he felt most inclined. In this there is postulated "an aetherial medium, much of the same constitution with air but far rarer, subtiler, and more strongly elastic." This aether is "a vibrating medium like air, only the vibrations far more swift and minute": and it "pervades the pores" of all natural bodies though standing at a "greater degree of rarity in those pores than in the free aetherial spaces." In passing from the Sun and Planets to the empty celestial spaces, it grows denser and denser perpetually, and thereby causes gravitation, "every body endeavouring to go from the denser parts of the medium towards the rarer."

The vibrations of the aether cannot, however, be supposed in themselves to constitute light, since the rectilinear propagation could not be explained thereby. "If it consisted in pression or motion, propagated either in an instant or in time, it would bend into the shadow." Light is therefore taken to be "something of a different kind, propagated from lucid bodies." Assuming then that the rays of light are "small bodies emitted every way from shining substances," these "where they impinge on any refracting or reflecting superficies, must as necessarily excite vibra-- tions in the aether, as stones do in water when thrown into it"; and by these vibrations light communicates heat to bodies. The aether is thus the intermediary between light and ponderable matter; for instance, in refraction a ray of light meets a stratum of aether denser or rarer than that through which it has lately been passing, and so is, in general deflected from its rectilinear course on account of the differences of density of the aether between one material medium and another. It is evident that aether-density plays much the same part in this hypothesis that the dielectric constant does in the electromagnetic theory of light: and Newton's suggestion regarding gravity resembles the modern hypothesis of Wiechert and his followers, that gravitational potential is an expression of what may be called the specific inductive capacity and permeability of the aether, these qualities being affected by the presence of gravitating bodies; and matter (assumed to be electrical in its nature) being attracted to places of greater dielectric constant.

It was assumed moreover that the condensation or rarefaction of the aether due to a material body extends to some little distance from the surface of the body, so that the inflexion due to it is really continuous, and not abrupt: and this further explains diffraction, which Newton took to be "only a new kind of refraction, caused, perhaps, by the external aether's beginning to grow rarer a little before it came at the opake body, than it was in free spaces."

The double part played by the aether in relation to light and gravitation led to a certain interlocking of these two subjects in Newton's mind: and when in the next decade he discovered that gravitation might be completely expressed by the law that every particle of matter in the universe attracts every other particle with a force which varies inversely as the square of the distance between them,—a law in which the aether is not mentioned—he began to regard the aether-hypothesis as something superfluous and to wish to get rid of it also from optics. A powerful argument against it now was that it could hardly be reconciled with his new Celestial Mechanics. "Against filling the Heavens with fluid mediums, unless they be exceeding rare, a great Objection arises from the regular and very lasting motions of the Planets and Comets. For thence it is manifest, that the Heavens are void of all sensible resistance, and by consequence of all sensible matter." Thus while Newton still kept up what one may call his official attitude of being independent of all hypotheses and indifferent to them, it may fairly be said that in the latter part of his life, he inclined more definitely to a purely corpuscular theory. He quotes with evident admiration "the oldest and most celebrated Philosophers of Greece and Phaenicia, who made a Vacuum, and Atoms, and the Gravity of Atoms, the first Principles of their Philosophy": and thus placed himself in the succession of Leucippus and Democritus. "It seems probable to me," he wrote in his last Query, "that God in the Beginning form'd Matter in solid, massy, hard, impenetrable, moveable Particles." Even the Rays of Light "seem to be hard Bodies," which are repelled by ordinary material bodies in the phenomena of reflexion and diffraction, the small particles of material bodies having certain "Powers, Virtues, or Forces, by which they act at a distance, not only upon the Rays of Light, but also upon one another for producing a great part of the Phenomena of Nature."

That Newton was more committed to the corpuscular theory than he realised is evident from the proof of the law of refraction which he gives in the sixth Proposition of the first Book of the *Optics*. He tells us that the proof is "general, without determining what Light is, or by what kind of Force it is refracted, or assuming any thing farther than that the refracting Body acts upon the Rays in lines perpendicular to its Surface": nevertheless on examining his proof, we find that it involves the proposition (afterwards repeated in Proposition ten of Book II) that the velocity of light is greater in a transparent medium than in a vacuum: which is a distinguishing mark of the corpuscular theory.

The inclination to the corpuscular theory in his later period is shown also by his treatment of two of the great problems of the subject: the colours of thin plates, and double refraction.

With regard to the colours of thin plates, a satisfactory explanation on undulatory principles might have been (and afterwards was) constructed by developing Hookes' rudimentary ideas of interference: Newton however took an entirely different line and invented the theory of *fits of easy reflexion and transmission* which is found in Proposition twelve of Book II of the *Optics*. He supposed that "every Ray of Light in its passage through any refracting Surface is put into a certain transient Constitution or State, which in the progress of the Ray returns at equal Intervals, and disposes the Ray at every return to be easily transmitted through the next refracting Surface, and between the returns to be easily reflected by it." The interval between two consecutive dispositions to easy transmission, or "length of fit," he supposed to depend on the colour, being greatest for red light and least for violet. If then a ray of homogeneous light falls on a thin plate, its fortunes as regards transmission and reflexion at the two surfaces will depend on the relation which the length of fit bears to the thickness of the plate: and on this basis he built up a theory of the colours of thin plates. It is evident that Newton's "length of fit" corresponds in some measure to the quantity which in the undulatory theory is called the wave-length of the light: but as he says, "What kind of action or disposition this is, whether it consists in a circulating or a vibrating motion of the Ray, or of the Medium, or something else, I do not here enquire," A somewhat vague explanation of the fits on undulatory principles is suggested in the seventeenth Query and further developed in Queries 28 and 29.

With regard to double refraction, it had been discovered by a Danish philosopher, Erasmus Bartholin, in 1669, and discussed by Huygens in his *Théorie de la lumière* in 1690. Newton took the matter up in the second edition of his *Optics*, published in 1717, and showed that Huygens' observations made it necessary to suppose that a ray obtained by double refraction differs from a ray of ordinary light in the same way as a long rod whose cross-section is a rectangle differs from a long rod whose cross-section is a circle: in other words, the properties of a ray of ordinary light are the same with respect to all directions at right angles to its direction of propagation, whereas a ray obtained by double refraction has properties related to special directions at right angles to its own direction. "Every Ray of Light," said Newton, "has therefore two opposite Sides, originally endowed with a Property on which the unusual Refraction depends, and the other two opposite sides not endued with that Property." The refraction of such a ray at the surface of a crystal depends on the relation of its sides to the principal plane of the crystal.

This brilliant intuition was the discovery of the *polarisation* of light—the name being taken from Newton's reference to the two poles of a magnet, in Query 29. Unfortunately this great advance led him, by a train of reasoning which was quite sound with respect to the knowledge of his day, to a still more decided rejection of the undulatory theory: that a ray of light should possess such properties seemed to him an insuperable objection to the hypothesis that light is constituted of waves analogous to waves of sound. "For Pressions or Motions," he says, "propagated from a shining Body through an uniform Medium, must be on all sides alike: whereas by those Experiments it appears, that the Rays of Light have different properties in their different Sides. . . . To me, at least, this seems inexplicable, if Light be nothing else than Pression or Motion propagated through Aether."

In this attitude he approached the theory of double refraction given in the *Théorie de la lumière*. Here Huygens gave the beautiful explanation of the refraction of the ordinary and extraordinary rays by means of the principle now known by his name, that the wave-front which represents the disturbance at any instant is the envelope of the secondary waves which arise from the various surface-elements of the wave-front at any preceding instant. But Newton was not in a frame of mind to appreciate anything which postulated waves in an aether, and in Query

25 he gives a rule for the refraction of the extra-ordinary ray which is actually not correct.

There remains only to say a word about the treatise before us. It was not his first attempt: for in February 1692 he left a light burning in his rooms when he went to chapel, "which, by unknown means, destroyed his papers, and among them a large work on Optics, containing the experiments and researches of twenty years." Its successor, the First edition of the present work, was not published till 1704: knowing as we do how Newton shrank from controversy, we may conjecture a possible connexion with the fact that his most pertinacious antagonist Hooke had been removed by death in 1703. Other editions, containing additional "Queries," appeared in 1717, 1721, and 1730; the last "corrected by the author's own hand, and left before his death with his bookseller": from it this reprint is taken.

E. T. WHITTAKER

ANALYTICAL TABLE OF CONTENTS*

THE FIRST BOOK OF OPTICKS
Book I, Part I

*Prepared by Duane H. D. Roller.

lxxix

THE SECOND BOOK OF OPTICKS

Book II, Part I

*Observations concerning the Reflexions,
Refractions, and Colours of thin
transparent Bodies*

Book II, Part II

Remarks upon the foregoing observations

Book II, Part III

Permanent colours of natural bodies and their analogy to colours of thin transparent plates

must be very porous—possibly the particles
are separated and themselves made up of
smaller, separated particles, which in turn
are made up of smaller, separated particles,
etc.—there are other possibilities—we do not
yet really know the inward frame of bodies.

heating of bodies—bending into the shadow does not occur with light as it does with water waves and sound—rays are bent passing edges but are bent away from the shadow and not towards it—double refraction cannot be explained by a wave theory—Huygens' attempts—difficulty of explaining double refraction by hypothesis of fits of easy reflexion and refraction—objection to filling the heavens with many fluid media—resistance of a fluid to motion through it due to both friction and inertia of the fluid—the inertia is the more important in such fluids as air, water, quicksilver, and is independent of size of fluid particles, depending only on density—motion in rarefied air—decrease in density with altitude—the observed planetary motion requires the heavens to be empty except for a slight amount of matter from planetary atmospheres and the aether—motions of planets and comets cannot be explained by means of a dense fluid and are better explained without it—there is no evidence for such a dense fluid medium and it should be rejected—this means rejecting the wave theory of light—for rejecting this medium we have the authority of Greek and Phoenician philosophers—the main business of natural philosophy is to argue from phaenomena without feigning hypotheses, to deduce causes from effects until the first cause is reached—it should deal with questions such as: What is in places almost devoid of matter? Why do bodies gravitate toward one another without dense matter between them? Whence is it that Nature does nothing in vain and whence arises all the order and beauty of the world? To what end are comets and why does their motion differ from that of the planets? Why don't the stars fall one upon another? How were animal bodies contrived with so much art and for what ends were their several parts? Was the eye contrived without skill in opticks and the ear without knowledge of sounds? How do the motions of the body follow from the will, and whence is the instinct in animals? Is not infinite space the sensory of a Being incorporeal, living, intelligent, omnipresent?—every true step in this direction brings us closer to a knowledge of the first cause, and on that account is to be highly valued.

362-370

Qu. 29. Rays of light are very small bodies emitted from shining substances—such bodies will have properties that conform to the phenomena—examples—the unusual refraction of Island-Crystal appears due to some attractive virtue lodged in certain sides of the rays and of the particles of the crystal—this virtue seems not magnetical, but is similar—it is difficult to conceive how rays of light can have a permanent virtue in two of their sides

and blood in motion and heat—if it were not
for such principles all motion, putrefaction,
generation, and life would cease—the probable state of matter at the creation and the
changes that have occurred since—in addition
to inertia the particles of matter have certain
active principles, which are not the occult
qualities of the Aristotelians—these occult
qualities put a stop to the improvement of
natural philosophy and tell us nothing—deriving general principles of motion from phenomena is a great step even if the causes of
those principles are not yet discovered—the
Creation—difficult things in natural philosophy should be investigated by the method of
analysis before being investigated by the
method of composition—this analysis consists in making experiments and observations
and in drawing general conclusions from them
by induction—hypotheses are not to be regarded in experimental philosophy—arguing
from experiments and observations by induction is the best way that the nature of things
admits of—proceeding in this way we may
proceed from effects to their causes, from
particular causes to more general ones, to the
most general—discussion of these Books of
Opticks in terms of this method—effect on
moral philosophy of perfecting natural philosophy—effect of worship of false gods on
the moral philosophy of the heathen   375-406

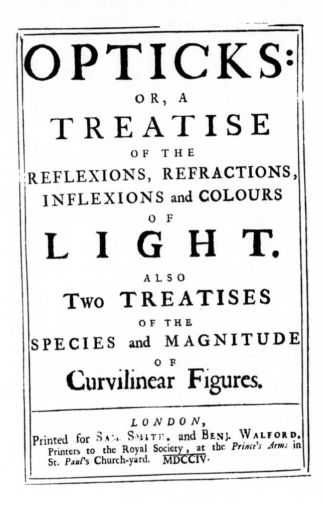

OPTICKS:

OR, A

TREATISE

OF THE

REFLEXIONS, REFRACTIONS, INFLEXIONS and COLOURS

OF

LIGHT.

ALSO

Two TREATISES

OF THE

SPECIES and MAGNITUDE

OF

Curvilinear Figures.

LONDON,
Printed for Sam. Smith. and Benj. Walford,
Printers to the Royal Society, at the *Prince's Arms* in
St. *Paul's* Church-yard. MDCCIV.

TITLE PAGE OF THE 1704 EDITION

OPTICKS:

OR, A

TREATISE

OF THE

Reflections, *Refractions*,
Inflections and *Colours*

OF

LIGHT.

The FOURTH EDITION, *corrected.*

By Sir *ISAAC NEWTON*, Knt.

LONDON:

Printed for WILLIAM INNYS at the West-
End of St. *Paul's.* MDCCXXX.

TITLE PAGE OF THE 1730 EDITION

SIR ISAAC NEWTON'S
ADVERTISEMENTS

Advertisement I

PART of the ensuing Discourse about Light was written at the Desire of some Gentlemen of the Royal-Society, in the Year 1675, and then sent to their Secretary, and read at their Meetings, and the rest was added about twelve Years after to complete the Theory; except the third Book, and the last Proposition of the Second, which were since put together out of scatter'd Papers. To avoid being engaged in Disputes about these Matters, I have hitherto delayed the printing, and should still have delayed it, had not the Importunity of Friends prevailed upon me. If any other Papers writ on this Subject are got out of my Hands they are imperfect, and were perhaps written before I had tried all the Experiments here set down, and fully satisfied my self about the Laws of Refractions and Composition of Colours. I have here publish'd what I think proper to come abroad, wishing that it may not be translated into another Language without my Consent.

The Crowns of Colours, which sometimes appear about the Sun and Moon, I have endeavoured to give an Account of; but for want of sufficient Observations leave that Matter to be farther examined. The Subject of the Third Book I have also left imperfect, not having tried all the Experiments which I intended when I was about these Matters, nor repeated some of those which I did try, until I had satisfied my self about all their Circumstances. To communicate what I have tried, and leave the rest to others for farther Enquiry, is all my Design in publishing these Papers.

In a Letter written to Mr. Leibnitz *in the year* 1679, *and published by Dr.* Wallis, *I mention'd a Method by which I had found some general Theorems about squaring Curvilinear Figures, or comparing them with the Conic Sections, or other the simplest Figures with which they may be compared. And some Years ago I lent out a Manuscript containing such Theorems, and having since met with some Things copied out of it, I have on this Occasion made it publick, prefixing to it an* Introduction, *and subjoining a* Scholium *concerning that Method. And I have joined with it another small Tract concerning the Curvilinear Figures of the Second Kind, which was also written many Years ago, and made known to some Friends, who have solicited the making it publick.*

April 1, 1704.   *I. N.*

Advertisement II

*I*N this Second Edition of these Opticks I have omitted the Mathematical Tracts publish'd at the End of the former Edition, as not belonging to the Subject. And at the End of the Third Book I have added some Questions. And to shew that I do not take Gravity for an essential Property of Bodies, I have added one Question concerning its Cause, chusing to propose it by way of a Question, because I am not yet satisfied about it for want of Experiments.

I. N.

July 16, 1717.

Advertisement to this Fourth Edition

*T*HIS new Edition of Sir Isaac Newton's Opticks is carefully printed from the Third Edition, as it was corrected by the Author's own Hand, and left before his Death with the Bookseller. Since Sir Isaac's Lectiones Opticæ, which he publickly read in the University of Cambridge in the Years 1669, 1670, and 1671, are lately printed, it has been thought proper to make at the bottom of the Pages several Citations from thence, where may be found the Demonstrations, which the Author omitted in these Opticks.

OPTICKS

THE
FIRST BOOK
OF
OPTICKS

PART I.

M Y Design in this Book is not to explain the Properties of Light by Hypotheses, but to propose and prove them by Reason and Experiments: In order to which I shall premise the following Definitions and Axioms.

DEFINITIONS
DEFIN. I.

B Y the Rays of Light I understand its least Parts, and those as well Successive in the same Lines, as Contemporary in several Lines. For it is manifest that Light consists of Parts, both Successive and Contemporary; because in the same place you may stop that which comes one moment, and let pass that which comes presently after; and in the same time you may stop it in any one place, and let it pass in any other. For that part of Light which is stopp'd cannot be the same with that which is let pass. The least Light or part of Light, which may be stopp'd alone without the rest of the Light, or propagated alone, or do or suffer any thing alone, which the rest of the Light doth not or suffers not, I call a Ray of Light.

DEFIN. II.

Refrangibility of the Rays of Light, is their Disposition to be refracted or turned out of their Way in passing out of one transparent Body or Medium into another. And a greater or less Refrangibility of Rays, is their Disposition to be turned more or less out of their Way in like Incidences on the same Medium. Mathematicians usually consider the Rays of Light to be Lines reaching from the luminous Body to the Body illuminated, and the refraction of those Rays to be the bending or breaking of those lines in their passing out of one Medium into another. And thus may Rays and Refractions be considered, if Light be propagated in an instant. But by an Argument taken from the Æquations of the times of the Eclipses of *Jupiter's Satellites*, it seems that Light is propagated in time, spending in its passage from the Sun to us about seven Minutes of time: And therefore I have chosen to define Rays and Refractions in such general terms as may agree to Light in both cases.

DEFIN. III.

Reflexibility of Rays, is their Disposition to be re-flected or turned back into the same Medium from any other Medium upon whose Surface they fall. And Rays are more or less reflexible, which are turned back more or less easily. As if Light pass out of a Glass into Air, and by being inclined more and more to the common Surface of the Glass and Air, begins at length to be totally reflected by that Surface; those sorts of Rays which at like Incidences are reflected most copiously, or by inclining the Rays begin soonest to be totally reflected, are most reflexible.

DEFIN. IV.

The Angle of Incidence is that Angle, which the Line described by the incident Ray contains with the Perpendicular to the reflecting or refracting Surface at the Point of Incidence.

DEFIN. V.

The Angle of Reflexion or Refraction, is the Angle which the line described by the reflected or refracted Ray containeth with the Perpendicular to the reflecting or refracting Surface at the Point of Incidence.

DEFIN. VI.

The Sines of Incidence, Reflexion, and Refraction, are the Sines of the Angles of Incidence, Reflexion, and Refraction.

DEFIN. VII.

The Light whose Rays are all alike Refrangible, I call Simple, Homogeneal and Similar; and that whose Rays are some more Refrangible than others, I call Compound, Heterogeneal and Dissimilar. The former Light I call Homogeneal, not because I would affirm it so in all respects, but because the Rays which agree in Refrangibility, agree at least in all those their other Properties which I consider in the following Discourse.

DEFIN. VIII.

The Colours of Homogeneal Lights, I call Primary, Homogeneal and Simple; and those of Heterogeneal Lights, Heterogeneal and Compound. For these are always compounded of the colours of Homogeneal Lights; as will appear in the following Discourse.

AXIOMS.

AX. I.

*T*HE *Angles of Reflexion and Refraction, lie in one and the same Plane with the Angle of Incidence.*

AX. II.

The Angle of Reflexion is equal to the Angle of Incidence.

AX. III.

If the refracted Ray be returned directly back to the Point of Incidence, it shall be refracted into the Line before described by the incident Ray.

AX. IV.

Refraction out of the rarer Medium into the denser, is made towards the Perpendicular; that is, so that the Angle of Refraction be less than the Angle of Incidence.

AX. V.

The Sine of Incidence is either accurately or very nearly in a given Ratio to the Sine of Refraction.

Whence if that Proportion be known in any one Inclination of the incident Ray, 'tis known in all the Inclinations, and thereby the Refraction in all cases of Incidence on the same refracting Body may be determined. Thus if the Refraction be made out of Air into Water, the Sine of Incidence of the red Light is to the Sine of its Refraction as 4 to 3. If out of Air into Glass, the Sines are as 17 to 11. In Light of other Colours the Sines have other Proportions: but the difference is so little that it need seldom be considered.

Suppose therefore, that RS [in *Fig.* 1.] represents the Surface of stagnating Water, and that C is the

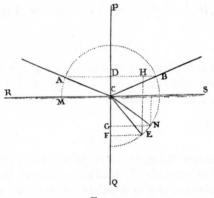

FIG. 1.

point of Incidence in which any Ray coming in the Air from A in the Line AC is reflected or refracted, and I would know whither this Ray shall go after Reflexion or Refraction: I erect upon the Surface of the Water from the point of Incidence the Perpendicular CP and produce it downwards to Q, and conclude by the first Axiom, that the Ray after Reflexion and

Refraction, shall be found somewhere in the Plane
of the Angle of Incidence ACP produced. I let fall
therefore upon the Perpendicular CP the Sine of Incidence AD; and if the reflected Ray be desired, I
produce AD to B so that DB be equal to AD, and
draw CB. For this Line CB shall be the reflected
Ray; the Angle of Reflexion BCP and its Sine BD
being equal to the Angle and Sine of Incidence, as
they ought to be by the second Axiom, But if the refracted Ray be desired, I produce AD to H, so that
DH may be to AD as the Sine of Refraction to the
Sine of Incidence, that is, (if the Light be red) as 3
to 4; and about the Center C and in the Plane ACP
with the Radius CA describing a Circle ABE, I draw
a parallel to the Perpendicular CPQ, the Line HE
cutting the Circumference in E, and joining CE, this
Line CE shall be the Line of the refracted Ray. For
if EF be let fall perpendicularly on the Line PQ, this
Line EF shall be the Sine of Refraction of the Ray
CE, the Angle of Refraction being ECQ ; and this
Sine EF is equal to DH, and consequently in Proportion to the Sine of Incidence AD as 3 to 4.

In like manner, if there be a Prism of Glass (that
is, a Glass bounded with two Equal and Parallel
Triangular ends, and three plain and well polished
Sides, which meet in three Parallel Lines running
from the three Angles of one end to the three Angles
of the other end) and if the Refraction of the Light
in passing cross this Prism be desired: Let ACB [in
Fig. 2.] represent a Plane cutting this Prism trans-
versly to its three Parallel lines or edges there where

the Light passeth through it, and let DE be the Ray
incident upon the first side of the Prism AC where
the Light goes into the Glass; and by putting the
Proportion of the Sine of Incidence to the Sine of
Refraction as 17 to 11 find EF the first refracted Ray.

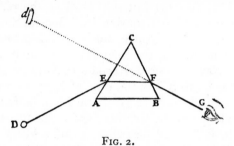

FIG. 2.

Then taking this Ray for the Incident Ray upon the
second side of the Glass BC where the Light goes
out, find the next refracted Ray FG by putting the
Proportion of the Sine of Incidence to the Sine of
Refraction as 11 to 17. For if the Sine of Incidence
out of Air into Glass be to the Sine of Refraction as
17 to 11, the Sine of Incidence out of Glass into Air
must on the contrary be to the Sine of Refraction as
11 to 17, by the third Axiom.

Much after the same manner, if ACBD [in *Fig.* 3.]
represent a Glass spherically convex on both sides
(usually called a *Lens*, such as is a Burning-glass, or
Spectacle-glass, or an Object-glass of a Telescope)
and it be required to know how Light falling upon it
from any lucid point Q shall be refracted, let QM represent a Ray falling upon any point M of its first

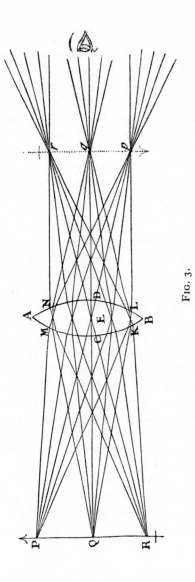

FIG. 3.

spherical Surface ACB, and by erecting a Perpendicular to the Glass at the point M, find the first refracted Ray MN by the Proportion of the Sines 17 to 11. Let that Ray in going out of the Glass be incident upon N, and then find the second refracted Ray N*q* by the Proportion of the Sines 11 to 17. And after the same manner may the Refraction be found when the Lens is convex on one side and plane or concave on the other, or concave on both sides.

AX. VI.

Homogeneal Rays which flow from several Points of any Object, and fall perpendicularly or almost perpendicularly on any reflecting or refracting Plane or spherical Surface, shall afterwards diverge from so many other Points, or be parallel to so many other Lines, or converge to so many other Points, either accurately or without any sensible Error. And the same thing will happen, if the Rays be reflected or refracted successively by two or three or more Plane or Spherical Surfaces.

The Point from which Rays diverge or to which they converge may be called their *Focus.* And the Focus of the incident Rays being given, that of the reflected or refracted ones may be found by finding the Refraction of any two Rays, as above; or more readily thus.

Cas. 1. Let ACB [in *Fig.* 4.] be a reflecting or refracting Plane, and Q the Focus of the incident Rays, and Q*q*C a Perpendicular to that Plane. And if this Perpendicular be produced to *q*, so that *q*C be equal

to QC, the Point q shall be the Focus of the reflected
Rays: Or if qC be taken on the same side of the Plane

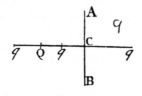

with QC, and in proportion to QC as the Sine of Incidence to the Sine of Refraction, the Point q shall be
the Focus of the refracted Rays.

 Cas. 2. Let ACB [in *Fig.* 5.] be the reflecting Surface of any Sphere whose Centre is E. Bisect any

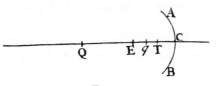

Radius thereof, (suppose EC) in T, and if in that
Radius on the same side the Point T you take the
Points Q and q, so that TQ, TE, and Tq, be continual
Proportionals, and the Point Q be the Focus of the
incident Rays, the Point q shall be the Focus of the
reflected ones.

 Cas. 3. Let ACB [in *Fig.* 6.] be the refracting Surface of any Sphere whose Centre is E. In any Radius
thereof EC produced both ways take ET and Ct

equal to one another and severally in such Proportion to that Radius as the lesser of the Sines of Incidence and Refraction hath to the difference of those Sines. And then if in the same Line you find any two Points

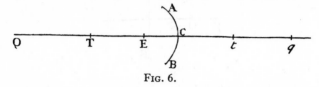

FIG. 6.

Q and q, so that TQ be to ET as Et to tq, taking tq the contrary way from t which TQ lieth from T, and if the Point Q be the Focus of any incident Rays, the Point q shall be the Focus of the refracted ones.

And by the same means the Focus of the Rays after two or more Reflexions or Refractions may be found.

Cas. 4. Let ACBD [in *Fig.* 7.] be any refracting Lens, spherically Convex or Concave or Plane on

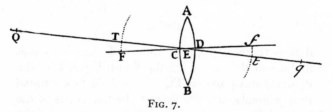

FIG. 7.

either side, and let CD be its Axis (that is, the Line which cuts both its Surfaces perpendicularly, and passes through the Centres of the Spheres,) and in this Axis produced let F and f be the Foci of the re-

fracted Rays found as above, when the incident Rays
on both sides the Lens are parallel to the same Axis;
and upon the Diameter F*f* bisected in E, describe a
Circle. Suppose now that any Point Q be the Focus
of any incident Rays. Draw QE cutting the said
Circle in T and *t*, and therein take *tq* in such proportion to *t*E as *t*E or TE hath to TQ. Let *tq* lie the contrary way from *t* which TQ doth from T, and *q* shall
be the Focus of the refracted Rays without any sensible Error, provided the Point Q be not so remote
from the Axis, nor the Lens so broad as to make any
of the Rays fall too obliquely on the refracting Surfaces.*

And by the like Operations may the reflecting or
refracting Surfaces be found when the two Foci are
given, and thereby a Lens be formed, which shall
make the Rays flow towards or from what Place you
please.†

So then the Meaning of this Axiom is, that if Rays
fall upon any Plane or Spherical Surface or Lens,
and before their Incidence flow from or towards any
Point Q, they shall after Reflexion or Refraction
flow from or towards the Point *q* found by the foregoing Rules. And if the incident Rays flow from or
towards several points Q, the reflected or refracted
Rays shall flow from or towards so many other Points

* In our Author's *Lectiones Opticæ*, Part I. Sect. IV. Prop 29,
30, there is an elegant Method of determining these *Foci*; not
only in spherical Surfaces, but likewise in any other curved
Figure whatever: And in Prop. 32, 33, the same thing is done
for any Ray lying out of the Axis.

† *Ibid*. Prop. 34.

q found by the same Rules. Whether the reflected and refracted Rays flow from or towards the Point *q* is easily known by the situation of that Point. For if that Point be on the same side of the reflecting or refracting Surface or Lens with the Point Q, and the incident Rays flow from the Point Q, the reflected flow towards the Point *q* and the refracted from it; and if the incident Rays flow towards Q, the reflected flow from *q*, and the refracted towards it. And the contrary happens when *q* is on the other side of the Surface.

AX. VII.

Wherever the Rays which come from all the Points of any Object meet again in so many Points after they have been made to converge by Reflection or Refraction, there they will make a Picture of the Object upon any white Body on which they fall.

So if PR [in *Fig.* 3.] represent any Object without Doors, and AB be a Lens placed at a hole in the Window-shut of a dark Chamber, whereby the Rays that come from any Point Q of that Object are made to converge and meet again in the Point *q*; and if a Sheet of white Paper be held at *q* for the Light there to fall upon it, the Picture of that Object PR will appear upon the Paper in its proper shape and Colours. For as the Light which comes from the Point Q goes to the Point *q*, so the Light which comes from other Points P and R of the Object, will go to so many other correspondent Points *p* and *r* (as is manifest by the sixth Axiom;) so that every Point of the Object shall illuminate a correspondent Point of the Picture, and thereby make a Picture like the Object in Shape and Colour, this only excepted, that the Picture shall be inverted. And this is the Reason of that vulgar Experiment of casting the Species of Objects from abroad upon a Wall or Sheet of white Paper in a dark Room.

In like manner, when a Man views any Object PQR, [in *Fig.* 8.] the Light which comes from the several Points of the Object is so refracted by the transparent skins and humours of the Eye, (that is, by the outward coat EFG, called the *Tunica Cornea*, and by the crystalline humour AB which is beyond the Pupil *mk*) as to converge and meet again in so many Points in the bottom of the Eye, and there to paint the Picture of the Object upon that skin (called the *Tunica Retina*) with which the bottom of the Eye is covered. For Anatomists, when they have taken off from the bottom of the Eye that outward and most thick Coat called the *Dura Mater*, can then see through the thinner Coats, the Pictures of Objects lively painted thereon. And these Pictures, propagated by Motion along the Fibres of the Optick Nerves into the Brain, are the cause of Vision. For accordingly as these Pictures are perfect or imperfect, the Object is seen perfectly or imperfectly. If the Eye be tinged with any colour (as in the Disease of the *Jaundice*) so as to tinge the Pictures in the bottom of the Eye with that Colour, then all Objects appear tinged with the same Colour. If the Humours of the Eye by old Age decay, so as by shrinking to

make the *Cornea* and Coat of the *Crystalline Humour*
grow flatter than before, the Light will not be refracted enough, and for want of a sufficient Refraction will not converge to the bottom of the Eye but
to some place beyond it, and by consequence paint in
the bottom of the Eye a confused Picture, and according to the Indistinctness of this Picture the
Object will appear confused. This is the reason of
the decay of sight in old Men, and shews why their
Sight is mended by Spectacles. For those Convex
glasses supply the defect of plumpness in the Eye,
and by increasing the Refraction make the Rays converge sooner, so as to convene distinctly at the
bottom of the Eye if the Glass have a due degree of
convexity. And the contrary happens in short-
sighted Meh whose Eyes are too plump. For the Refraction being now too great, the Rays converge and
convene in the Eyes before they come at the bottom;
and therefore the Picture made in the bottom and the
Vision caused thereby will not be distinct, unless the
Object be brought so near the Eye as that the place
where the converging Rays convene may be removed
to the bottom, or that the plumpness of the Eye be
taken off and the Refractions diminished by a Con-
cave-glass of a due degree of Concavity, or lastly
that by Age the Eye grow flatter till it come to a due
Figure: For short-sighted Men see remote Objects
best in Old Age, and therefore they are accounted to
have the most lasting Eyes.

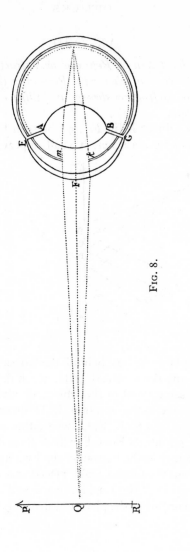

Fig. 8.

AX. VIII.

*An Object seen by Reflexion or Refraction, appears
in that place from whence the Rays after their last Reflexion or Refraction diverge in falling on the Spectator's Eye.*

If the Object A [in *Fig.* 9.] be seen by Reflexion
of a Looking-glass *mn*, it shall appear, not in its

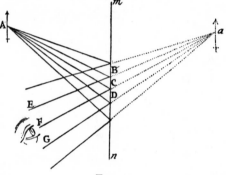

FIG. 9.

proper place A, but behind the Glass at *a*, from
whence any Rays AB, AC, AD, which flow from one
and the same Point of the Object, do after their Reflexion made in the Points B, C, D, diverge in going
from the Glass to E, F, G, where they are incident
on the Spectator's Eyes. For these Rays do make the
same Picture in the bottom of the Eyes as if they had
come from the Object really placed at *a* without the
Interposition of the Looking-glass; and all Vision is
made according to the place and shape of that Picture.

In like manner the Object D [in *Fig.* 2.] seen through a Prism, appears not in its proper place D, but is thence translated to some other place *d* situated in the last refracted Ray FG drawn backward from F to *d*.

And so the Object Q [in *Fig.* 10.] seen through the Lens AB, appears at the place *q* from whence the

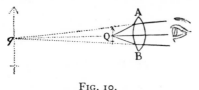

FIG. 10.

Rays diverge in passing from the Lens to the Eye. Now it is to be noted, that the Image of the Object at *q* is so much bigger or lesser than the Object it self at Q, as the distance of the Image at *q* from the Lens AB is bigger or less than the distance of the Object at Q from the same Lens. And if the Object be seen through two or more such Convex or Concave-glasses, every Glass shall make a new Image, and the Object shall appear in the place of the bigness of the last Image. Which consideration unfolds the Theory of Microscopes and Telescopes. For that Theory consists in almost nothing else than the describing such Glasses as shall make the last Image of any Object as distinct and large and luminous as it can conveniently be made.

I have now given in Axioms and their Explications the sum of what hath hitherto been treated of in

Opticks. For what hath been generally agreed on I content my self to assume under the notion of Principles, in order to what I have farther to write. And this may suffice for an Introduction to Readers of quick Wit and good Understanding not yet versed in Opticks: Although those who are already acquainted with this Science, and have handled Glasses, will more readily apprehend what followeth.

PROPOSITIONS.

PROP. I. THEOR. I.

Lights which differ in Colour, differ also in Degrees of Refrangibility.

The PROOF by Experiments.

Exper. 1. I took a black oblong stiff Paper terminated by Parallel Sides, and with a Perpendicular right Line drawn cross from one Side to the other, distinguished it into two equal Parts. One of these parts I painted with a red colour and the other with a blue. The Paper was very black, and the Colours intense and thickly laid on, that the Phænomenon might be more conspicuous. This Paper I view'd through a Prism of solid Glass, whose two Sides through which the Light passed to the Eye were plane and well polished, and contained an Angle of about sixty degrees; which Angle I call the refracting Angle of the Prism. And whilst I view'd it,

I held it and the Prism before a Window in such manner that the Sides of the Paper were parallel to the Prism, and both those Sides and the Prism were parallel to the Horizon, and the cross Line was also parallel to it: and that the Light which fell from the Window upon the Paper made an Angle with the Paper, equal to that Angle which was made with the same Paper by the Light reflected from it to the Eye. Beyond the Prism was the Wall of the Chamber under the Window covered over with black Cloth, and the Cloth was involved in Darkness that no Light might be reflected from thence, which in passing by the Edges of the Paper to the Eye, might mingle itself with the Light of the Paper, and obscure the Phænomenon thereof. These things being thus ordered, I found that if the refracting Angle of the Prism be turned upwards, so that the Paper may seem to be lifted upwards by the Refraction, its blue half will be lifted higher by the Refraction than its red half. But if the refracting Angle of the Prism be turned downward, so that the Paper may seem to be carried lower by the Refraction, its blue half will be carried something lower thereby than its red half. Wherefore in both Cases the Light which comes from the blue half of the Paper through the Prism to the Eye, does in like Circumstances suffer a greater Refraction than the Light which comes from the red half, and by consequence is more refrangible.

Illustration. In the eleventh Figure, MN represents the Window, and DE the Paper terminated with parallel Sides DJ and HE, and by the transverse

Line FG distinguished into two halfs, the one DG of an intensely blue Colour, the other FE of an intensely red. And BAC*cab* represents the Prism whose

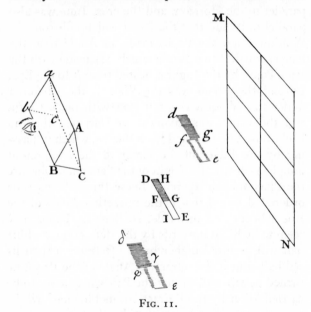

FIG. 11.

refracting Planes AB*ba* and AC*ca* meet in the Edge of the refracting Angle *Aa*. This Edge A*a* being upward, is parallel both to the Horizon, and to the Parallel-Edges of the Paper DJ and HE, and the transverse Line FG is perpendicular to the Plane of the Window. And *de* represents the Image of the Paper seen by Refraction upwards in such manner, that the blue half DG is carried higher to *dg* than the

red half FE is to *fe*, and therefore suffers a greater
Refraction. If the Edge of the refracting Angle be
turned downward, the Image of the Paper will be
refracted downward; suppose to $\delta\varepsilon$, and the blue
half will be refracted lower to $\delta\gamma$ than the red half
is to $\varphi\varepsilon$.

Exper. 2. About the aforesaid Paper, whose two
halfs were painted over with red and blue, and which
was stiff like thin Pasteboard, I lapped several times
a slender Thred of very black Silk, in such manner
that the several parts of the Thred might appear
upon the Colours like so many black Lines drawn
over them, or like long and slender dark Shadows
cast upon them. I might have drawn black Lines
with a Pen, but the Threds were smaller and better
defined. This Paper thus coloured and lined I set
against a Wall perpendicularly to the Horizon, so
that one of the Colours might stand to the Right
Hand, and the other to the Left. Close before the
Paper, at the Confine of the Colours below, I placed
a Candle to illuminate the Paper strongly: For the
Experiment was tried in the Night. The Flame of
the Candle reached up to the lower edge of the
Paper, or a very little higher. Then at the distance of
six Feet, and one or two Inches from the Paper upon
the Floor I erected a Glass Lens four Inches and a
quarter broad, which might collect the Rays coming
from the several Points of the Paper, and make them
converge towards so many other Points at the same
distance of six Feet, and one or two Inches on the
other side of the Lens, and so form the Image of the

coloured Paper upon a white Paper placed there, after the same manner that a Lens at a Hole in a Window casts the Images of Objects abroad upon a Sheet of white Paper in a dark Room. The aforesaid white Paper, erected perpendicular to the Horizon, and to the Rays which fell upon it from the Lens, I moved sometimes towards the Lens, sometimes from it, to find the Places where the Images of the blue and red Parts of the coloured Paper appeared most distinct. Those Places I easily knew by the Images of the black Lines which I had made by winding the Silk about the Paper. For the Images of those fine and slender Lines (which by reason of their Blackness were like Shadows on the Colours) were confused and scarce visible, unless when the Colours on either side of each Line were terminated most distinctly, Noting therefore, as diligently as I could, the Places where the Images of the red and blue halfs of the coloured Paper appeared most distinct, I found that where the red half of the Paper appeared distinct, the blue half appeared confused, so that the black Lines drawn upon it could scarce be seen; and on the contrary, where the blue half appeared most distinct, the red half appeared confused, so that the black Lines upon it were scarce visible. And between the two Places where these Images appeared distinct there was the distance of an Inch and a half; the distance of the white Paper from the Lens, when the Image of the red half of the coloured Paper appeared most distinct, being greater by an Inch and an half than the distance of the same white Paper from the

Lens, when the Image of the blue half appeared most distinct. In like Incidences therefore of the blue and red upon the Lens, the blue was refracted more by the Lens than the red, so as to converge sooner by an Inch and a half, and therefore is more refrangible.

Illustration. In the twelfth Figure (p. 27), DE signifies the coloured Paper, DG the blue half, FE the red half, MN the Lens, HJ the white Paper in that Place where the red half with its black Lines appeared distinct, and *hi* the same Paper in that Place where the blue half appeared distinct. The Place *hi* was nearer to the Lens MN than the Place HJ by an Inch and an half.

Scholium. The same Things succeed, notwithstanding that some of the Circumstances be varied; as in the first Experiment when the Prism and Paper are any ways inclined to the Horizon, and in both when coloured Lines are drawn upon very black Paper. But in the Description of these Experiments, I have set down such Circumstances, by which either the Phænomenon might be render'd more conspicuous, or a Novice might more easily try them, or by which I did try them only. The same Thing, I have often done in the following Experiments: Concerning all which, this one Admonition may suffice. Now from these Experiments it follows not, that all the Light of the blue is more refrangible than all the Light of the red: For both Lights are mixed of Rays differently refrangible, so that in the red there are some Rays not less refrangible than those of the blue,

and in the blue there are some Rays not more refrangible than those of the red: But these Rays, in
proportion to the whole Light, are but few, and serve
to diminish the Event of the Experiment, but are not
able to destroy it. For, if the red and blue·Colours
were more dilute and weak, the distance of the
Images would be less than an Inch and a half; and if
they were more intense and full, that distance would
be greater, as will appear hereafter. These Experiments may suffice for the Colours of Natural Bodies.
For in the Colours made by the Refraction of Prisms,
this Proposition will appear by the Experiments
which are now to follow in the next Proposition.

PROP. II.   THEOR. II.

*The Light of the Sun consists of Rays differently
Refrangible.*

The PROOF by Experiments.

Exper. 3.   IN a very dark Chamber, at a round Hole,
about one third Part of an Inch broad,
made in the Shut of a Window, I placed a Glass
Prism, whereby the Beam of the Sun's Light, which
came in at that Hole, might be refracted upwards
toward the opposite Wall of the Chamber, and there
form a colour'd Image of the Sun. The Axis of the
Prism (that is, the Line passing through the middle
of the Prism from one end of it to the other end
parallel to the edge of the Refracting Angle) was in

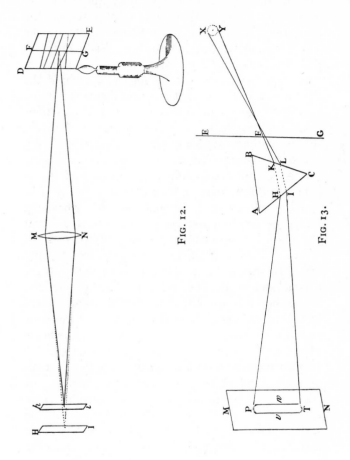

FIG. 12.

FIG. 13.

this and the following Experiments perpendicular to
the incident Rays. About this Axis I turned the
Prism slowly, and saw the refracted Light on the
Wall, or coloured Image of the Sun, first to descend,
and then to ascend. Between the Descent and Ascent,
when the Image seemed Stationary, I stopp'd the
Prism, and fix'd it in that Posture, that it should be
moved no more. For in that Posture the Refractions
of the Light at the two Sides of the refracting Angle,
that is, at the Entrance of the Rays into the Prism,
and at their going out of it,were equal to one another.*
So also in other Experiments, as often as I would
have the Refractions on both sides the Prism to be
equal to one another, I noted the Place where the
Image of the Sun formed by the refracted Light
stood still between its two contrary Motions, in the
common Period of its Progress and Regress; and
when the Image fell upon that Place, I made fast the
Prism. And in this Posture, as the most convenient,
it is to be understood that all the Prisms are placed
in the following Experiments, unless where some
other Posture is described. The Prism therefore being
placed in this Posture, I let the refracted Light fall
perpendicularly upon a Sheet of white Paper at the
opposite Wall of the Chamber, and observed the
Figure and Dimensions of the Solar Image formed
on the Paper by that Light. This Image was Oblong
and not Oval, but terminated with two Rectilinear
and Parallel Sides, and two Semicircular Ends. On

* See our Author's Lectiones Opticæ, Part I. Sect. 1. § 10.
Sect. II. § 29. and Sect. III. Prop. 25.

its Sides it was bounded pretty distinctly, but on its
Ends very confusedly and indistinctly, the Light
there decaying and vanishing by degrees. The
Breadth of this Image answered to the Sun's Diameter, and was about two Inches and the eighth Part
of an Inch, including the Penumbra. For the Image
was eighteen Feet and an half distant from the
Prism, and at this distance that Breadth, if diminished
by the Diameter of the Hole in the Window-shut,
that is by a quarter of an Inch, subtended an Angle
at the Prism of about half a Degree, which is the
Sun's apparent Diameter. But the Length of the
Image was about ten Inches and a quarter, and the
Length of the Rectilinear Sides about eight Inches;
and the refracting Angle of the Prism, whereby so
great a Length was made, was 64 degrees. With a less
Angle the Length of the Image was less, the Breadth
remaining the same. If the Prism was turned about
its Axis that way which made the Rays emerge more
obliquely out of the second refracting Surface of the
Prism, the Image soon became an Inch or two longer,
or more; and if the Prism was turned about the contrary way, so as to make the Rays fall more obliquely
on the first refracting Surface, the Image soon became an Inch or two shorter. And therefore in trying
this Experiment, I was as curious as I could be in
placing the Prism by the above-mention'd Rule exactly in such a Posture, that the Refractions of the
Rays at their Emergence out of the Prism might be
equal to that at their Incidence on it. This Prism had
some Veins running along within the Glass from one

end to the other, which scattered some of the Sun's
Light irregularly, but had no sensible Effect in increasing the Length of the coloured Spectrum. For
I tried the same Experiment with other Prisms with
the same Success. And particularly with a Prism
which seemed free from such Veins, and whose refracting Angle was $62\frac{1}{2}$ Degrees, I found the Length
of the Image $9\frac{3}{4}$ or 10 Inches at the distance of $18\frac{1}{2}$
Feet from the Prism, the Breadth of the Hole in the
Window-shut being $\frac{1}{4}$ of an Inch, as before. And
because it is easy to commit a Mistake in placing the
Prism in its due Posture, I repeated the Experiment
four or five Times, and always found the Length of
the Image that which is set down above. With
another Prism of clearer Glass and better Polish,
which seemed free from Veins, and whose refracting
Angle was $63\frac{1}{2}$ Degrees, the Length of this Image at
the same distance of $18\frac{1}{2}$ Feet was also about 10
Inches, or $10\frac{1}{8}$. Beyond these Measures for about a
$\frac{1}{4}$ or $\frac{1}{3}$ of an Inch at either end of the Spectrum the
Light of the Clouds seemed to be a little tinged with
red and violet, but so very faintly, that I suspected
that Tincture might either wholly, or in great
Measure arise from some Rays of the Spectrum
scattered irregularly by some Inequalities in the
Substance and Polish of the Glass, and therefore I
did not include it in these Measures. Now the
different Magnitude of the hole in the Window-shut,
and different thickness of the Prism where the Rays
passed through it, and different inclinations of the
Prism to the Horizon, made no sensible changes in

the length of the Image. Neither did the different
matter of the Prisms make any : for in a Vessel made
of polished Plates of Glass cemented together in the
shape of a Prism and filled with Water, there is the
like Success of the Experiment according to the
quantity of the Refraction. It is farther to be observed, that the Rays went on in right Lines from
the Prism to the Image, and therefore at their very
going out of the Prism had all that Inclination to one
another from which the length of the Image proceeded, that is, the Inclination of more than two
degrees and an half. And yet according to the Laws
of Opticks vulgarly received, they could not possibly
be so much inclined to one another.* For let EG [*Fig.*
13. (p. 27)] represent the Window-shut, F the hole
made therein through which a beam of the Sun's
Light was transmitted into the darkened Chamber,
and ABC a Triangular Imaginary Plane whereby the
Prism is feigned to be cut transversely through the
middle of the Light. Or if you please, let ABC represent the Prism it self, looking directly towards the
Spectator's Eye with its nearer end: And let XY be
the Sun, MN the Paper upon which the Solar
Image or Spectrum is cast, and PT the Image it self
whose sides towards *v* and *w* are Rectilinear and
Parallel, and ends towards P and T Semicircular.
YKHP and XLJT are two Rays, the first of which
comes from the lower part of the Sun to the higher
part of the Image, and is refracted in the Prism at K
and H, and the latter comes from the higher part of

* See our Author's *Lectiones Opticæ*, Part. I. Sect. 1. § 5.

the Sun to the lower part of the Image, and is refracted at L and J. Since the Refractions on both
sides the Prism are equal to one another, that is, the
Refraction at K equal to the Refraction at J, and the
Refraction at L equal to the Refraction at H, so that
the Refractions of the incident Rays at K and L
taken together, are equal to the Refractions of the
emergent Rays at H and J taken together: it follows
by adding equal things to equal things, that the Refractions at K and H taken together, are equal to the
Refractions at J and L taken together, and therefore
the two Rays being equally refracted, have the same
Inclination to one another after Refraction which
they had before; that is, the Inclination of half a
Degree answering to the Sun's Diameter. For so
great was the inclination of the Rays to one another
before Refraction. So then, the length of the Image
PT would by the Rules of Vulgar Opticks subtend
an Angle of half a Degree at the Prism, and by Consequence be equal to the breadth vw; and therefore
the Image would be round. Thus it would be were
the two Rays XLJT and YKHP, and all the rest
which form the Image PwTv, alike refrangible. And
therefore seeing by Experience it is found that the
Image is not round, but about five times longer than
broad, the Rays which going to the upper end P of
the Image suffer the greatest Refraction, must be
more refrangible than those which go to the lower
end T, unless the Inequality of Refraction be casual.

This Image or Spectrum PT was coloured, being
red at its least refracted end T, and violet at its most

refracted end P, and yellow green and blue in the
intermediate Spaces. Which agrees with the first
Proposition, that Lights which differ in Colour, do
also differ in Refrangibility. The length of the Image
in the foregoing Experiments, I measured from the
faintest and outmost red at one end, to the faintest
and outmost blue at the other end, excepting only
a little Penumbra, whose breadth scarce exceeded a
quarter of an Inch, as was said above.

Exper. 4. In the Sun's Beam which was propagated into the Room through the hole in the Window-
shut, at the distance of some Feet from the hole, I
held the Prism in such a Posture, that its Axis might
be perpendicular to that Beam. Then I looked
through the Prism upon the hole, and turning the
Prism to and fro about its Axis, to make the Image of
the Hole ascend and descend, when between its two
contrary Motions it seemed Stationary, I stopp'd the
Prism, that the Refractions of both sides of the refracting Angle might be equal to each other, as in the
former Experiment. In this situation of the Prism
viewing through it the said Hole, I observed the
length of its refracted Image to be many times
greater than its breadth, and that the most refracted
part thereof appeared violet, the least refracted red,
the middle parts blue, green and yellow in order.
The same thing happen'd when I removed the
Prism out of the Sun's Light, and looked through it
upon the hole shining by the Light of the Clouds
beyond it. And yet if the Refraction were done
regularly according to one certain Proportion of the

Sines of Incidence and Refraction as is vulgarly
supposed, the refracted Image ought to have
appeared round.

So then, by these two Experiments it appears,
that in Equal Incidences there is a considerable inequality of Refractions. But whence this inequality
arises, whether it be that some of the incident Rays
are refracted more, and others less, constantly, or
by chance, or that one and the same Ray is by Refraction disturbed, shatter'd, dilated, and as it were
split and spread into many diverging Rays, as *Grimaldo* supposes, does not yet appear by these Experiments, but will appear by those that follow.

Exper 5. Considering therefore, that if in the third
Experiment the Image of the Sun should be drawn
out into an oblong Form, either by a Dilatation of
every Ray, or by any other casual inequality of the
Refractions, the same oblong Image would by a
second Refraction made sideways be drawn out as
much in breadth by the like Dilatation of the Rays,
or other casual inequality of the Refractions sideways, I tried what would be the Effects of such a
second Refraction. For this end I ordered all things
as in the third Experiment, and then placed a second
Prism immediately after the first in a cross Position
to it, that it might again refract the beam of the Sun's
Light which came to it through the first Prism.   In
the first Prism this beam was refracted upwards, and
in the second sideways. And I found that by the Refraction of the second Prism, the breadth of the
Image was not increased, but its superior part,

which in the first Prism suffered the greater Refraction, and appeared violet and blue, did again in the
second Prism suffer a greater Refraction than its
inferior part, which appeared red and yellow, and
this without any Dilatation of the Image in breadth.

   Illustration. Let S [*Fig.* 14, 15.] represent the Sun,
F the hole in the Window, ABC the first Prism, DH
the second Prism, Y the round Image of the Sun made
by a direct beam of Light when the Prisms are taken
away, PT the oblong Image of the Sun made by that
beam passing through the first Prism alone, when
the second Prism is taken away, and *pt* the Image
made by the cross Refractions of both Prisms together. Now if the Rays which tend towards the
several Points of the round Image Y were dilated
and spread by the Refraction of the first Prism, so
that they should not any longer go in single Lines
to single Points, but that every Ray being split,
shattered, and changed from a Linear Ray to a
Superficies of Rays diverging from the Point of Refraction, and lying in the Plane of the Angles of Incidence and Refraction, they should go in those Planes
to so many Lines reaching almost from one end of
the Image PT to the other, and if that Image should
thence become oblong: those Rays and their several
parts tending towards the several Points of the
Image PT ought to be again dilated and spread sideways by the transverse Refraction of the second
Prism, so as to compose a four square Image, such as
is represented at $\pi\tau$. For the better understanding of
which, let the Image PT be distinguished into five

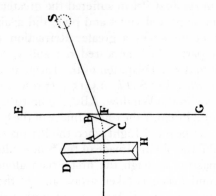

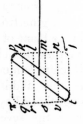

FIG. 14

equal parts PQK, KQRL, LRSM, MSVN, NVT. And by the same irregularity that the orbicular Light Y is by the Refraction of the first Prism dilated and drawn out into a long Image PT, the Light PQK which takes up a space of the same length and breadth with the Light Y ought to be by the Refraction of the second Prism dilated and drawn out into the long Image πqkp, and the Light KQRL into the long Image kqrl, and the Lights LRSM, MSVN, NVT, into so many other long Images lrsm, msvn, nvtτ; and all these long Images would compose the four square Images πτ. Thus it ought to be were every Ray dilated by Refraction, and spread into a triangular Superficies of Rays diverging from the Point of Refraction. For the second Refraction would spread the Rays one way as much as the first doth another, and so dilate the Image in breadth as much as the first doth in length. And the same thing ought to happen, were some rays casually refracted more than others. But the Event is otherwise. The Image PT was not made broader by the Refraction of the second Prism, but only became oblique, as 'tis represented at pt, its upper end P being by the Refraction translated to a greater distance than its lower end T. So then the Light which went towards the upper end P of the Image, was (at equal Incidences) more refracted in the second Prism, than the Light which tended towards the lower end T, that is the blue and violet, than the red and yellow; and therefore was more refrangible. The same Light was by the Refraction of the first Prism translated farther from the place Y to which it tended before Refraction; and therefore suffered as well in the first Prism as in the second a greater Refraction than the rest of the Light, and by consequence was more refrangible than the rest, even before its incidence on the first Prism.

Sometimes I placed a third Prism after the second, and sometimes also a fourth after the third, by all which the Image might be often refracted sideways: but the Rays which were more refracted than the rest in the first Prism were also more refracted in all the rest, and that without any Dilatation of the Image sideways : and therefore those Rays for their constancy of a greater Refraction are deservedly reputed more refrangible.

But that the meaning of this Experiment may more clearly appear, it is to be considered that the Rays which are equally refrangible do fall upon a Circle answering to the Sun's Disque. For this was proved in the third Experiment. By a Circle I understand not here a perfect geometrical Circle, but any orbicular Figure whose length is equal to its breadth, and which, as to Sense, may seem circular. Let therefore AG [in *Fig.* 15.] represent the Circle which all the most refrangible Rays propagated from the whole Disque of the Sun, would illuminate and paint upon the opposite Wall if they were alone; EL the Circle which all the least refrangible Rays would in like manner illuminate and paint if they were alone; BH, CJ, DK, the Circles which so many intermediate sorts of Rays would successively paint upon the

Wall, if they were singly propagated from the Sun in successive order, the rest being always intercepted; and conceive that there are other intermediate Circles without Number, which innumerable other intermediate sorts of Rays would successively paint upon the Wall if the Sun should successively emit

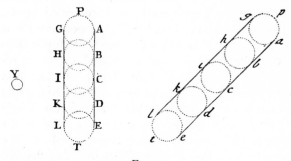

FIG. 15

every sort apart. And seeing the Sun emits all these sorts at once, they must all together illuminate and paint innumerable equal Circles, of all which, being according to their degrees of Refrangibility placed in order in a continual Series, that oblong Spectrum PT is composed which I described in the third Experiment. Now if the Sun's circular Image Y [in *Fig.* 15.] which is made by an unrefracted beam of Light was by any Dilation of the single Rays, or by any other irregularity in the Refraction of the first Prism, converted into the oblong Spectrum, PT: then ought every Circle AG, BH, CJ, &c. in that Spectrum, by the cross Refraction of the second

Prism again dilating or otherwise scattering the Rays as before, to be in like manner drawn out and transformed into an oblong Figure, and thereby the breadth of the Image PT would be now as much augmented as the length of the Image Y was before by the Refraction of the first Prism; and thus by the Refractions of both Prisms together would be formed a four square Figure $p\pi t\tau$, as I described above. Wherefore since the breadth of the Spectrum PT is not increased by the Refraction sideways, it is certain that the Rays are not split or dilated, or otherways irregularly scatter'd by that Refraction, but that every Circle is by a regular and uniform Refraction translated entire into another Place, as the Circle AG by the greatest Refraction into the place ag, the Circle BH by a less Refraction into the place bh, the Circle CJ by a Refraction still less into the place ci, and so of the rest; by which means a new Spectrum pt inclined to the former PT is in like manner composed of Circles lying in a right Line; and these Circles must be of the same bigness with the former, because the breadths of all the Spectrums Y, PT and pt at equal distances from the Prisms are equal.

I considered farther, that by the breadth of the hole F through which the Light enters into the dark Chamber, there is a Penumbra made in the Circuit of the Spectrum Y, and that Penumbra remains in the rectilinear Sides of the Spectrums PT and pt. I placed therefore at that hole a Lens or Object-glass of a Telescope which might cast the Image of the Sun distinctly on Y without any Penumbra at all,

and found that the Penumbra of the rectilinear Sides
of the oblong Spectrums PT and *pt* was also thereby
taken away, so that those Sides appeared as distinctly defined as did the Circumference of the first
Image Y. Thus it happens if the Glass of the Prisms
be free from Veins, and their sides be accurately
plane and well polished without those numberless
Waves or Curles which usually arise from Sand-
holes a little smoothed in polishing with Putty. If
the Glass be only well polished and free from Veins,
and the Sides not accurately plane, but a little Convex or Concave, as it frequently happens ; yet may
the three Spectrums Y, PT and *pt* want Penumbras,
but not in equal distances from the Prisms. Now
from this want of Penumbras, I knew more certainly
that every one of the Circles was refracted according
to some most regular, uniform and constant Law.
For if there were any irregularity in the Refraction,
the right Lines AE and GL, which all the Circles
in the Spectrum PT do touch, could not by that Refraction be translated into the Lines *ae* and *gl* as
distinct and straight as they were before, but there
would arise in those translated Lines some Penumbra or Crookedness or Undulation, or other
sensible Perturbation contrary to what is found by
Experience. Whatsoever Penumbra or Perturbation
should be made in the Circles by the cross Refraction of the second Prism, all that Penumbra or Perturbation would be conspicuous in the right Lines
ae and *gl* which touch those Circles. And therefore
since there is no such Penumbra or Perturbation in

those right Lines, there must be none in the Circles. Since the distance between those Tangents or breadth of the Spectrum is not increased by the Refractions, the Diameters of the Circles are not increased thereby. Since those Tangents continue to be right Lines, every Circle which in the first Prism is more or less refracted, is exactly in the same proportion more or less refracted in the second. And seeing all these things continue to succeed after the same manner when the Rays are again in a third Prism, and again in a fourth refracted sideways, it is evident that the Rays of one and the same Circle, as to their degree of Refrangibility, continue always uniform and homogeneal to one another, and that those of several Circles do differ in degree of Refrangibility, and that in some certain and constant Proportion. Which is the thing I was to prove.

There is yet another Circumstance or two of this Experiment by which it becomes still more plain and convincing. Let the second Prism DH [in *Fig.* 16.] be placed not immediately after the first, but at some distance from it; suppose in the mid-way between it and the Wall on which the oblong Spectrum PT is cast, so that the Light from the first Prism may fall upon it in the form of an oblong Spectrum $\pi\tau$ parallel to this second Prism, and be refracted sideways to form the oblong Spectrum pt upon the Wall. And you will find as before, that this Spectrum pt is inclined to that Spectrum PT, which the first Prism forms alone without the second; the blue ends P and p being farther distant from one another than the

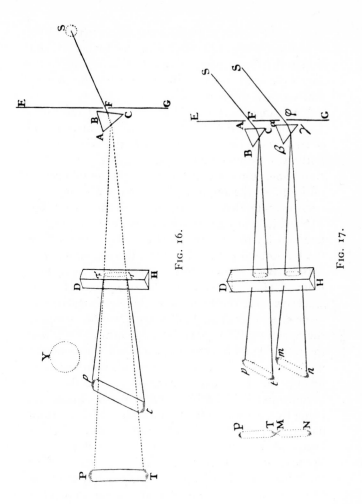

Fig. 16.

Fig. 17.

red ones T and *t*, and by consequence that the Rays which go to the blue end π of the Image πτ, and which therefore suffer the greatest Refraction in the first Prism, are again in the second Prism more refracted than the rest.

The same thing I try'd also by letting the Sun's Light into a dark Room through two little round holes F and φ [in *Fig.* 17.] made in the Window, and with two parallel Prisms ABC and αβγ placed at those holes (one at each) refracting those two beams of Light to the opposite Wall of the Chamber, in such manner that the two colour'd Images PT and MN which they there painted were joined end to end and lay in one straight Line, the red end T of the one touching the blue end M of the other. For if these two refracted Beams were again by a third Prism DH placed cross to the two first, refracted sideways, and the Spectrums thereby translated to some other part of the Wall of the Chamber, suppose the Spectrum PT to *pt* and the Spectrum MN to *mn*, these translated Spectrums *pt* and *mn* would not lie in one straight Line with their ends contiguous as before, but be broken off from one another and become parallel, the blue end *m* of the Image *mn* being by a greater Refraction translated farther from its former place MT, than the red end *t* of the other Image *pt* from the same place MT; which puts the Proposition past Dispute. And this happens whether the third Prism DH be placed immediately after the two first, or at a great distance from them, so that the Light refracted in the two first Prisms be either

white and circular, or coloured and oblong when it
falls on the third.

Exper. 6. In the middle of two thin Boards I made
round holes a third part of an Inch in diameter, and
in the Window-shut a much broader hole being made
to let into my darkned Chamber a large Beam of the
Sun's Light; I placed a Prism behind the Shut in
that beam to refract it towards the opposite Wall, and
close behind the Prism I fixed one of the Boards,
in such manner that the middle of the refracted
Light might pass through the hole made in it, and
the rest be intercepted by the Board. Then at the
distance of about twelve Feet from the first Board I
fixed the other Board in such manner that the middle
of the refracted Light which came through the hole
in the first Board, and fell upon the opposite Wall,
might pass through the hole in this other Board, and
the rest being intercepted by the Board might paint
upon it the coloured Spectrum of the Sun. And close
behind this Board I fixed another Prism to refract
the Light which came through the hole. Then I returned speedily to the first Prism, and by turning it
slowly to and fro about its Axis, I caused the Image
which fell upon the second Board to move up and
down upon that Board, that all its parts might
successively pass through the hole in that Board and
fall upon the Prism behind it. And in the mean time,
I noted the places on the opposite Wall to which that
Light after its Refraction in the second Prism did
pass; and by the difference of the places I found that
the Light which being most refracted in the first

Prism did go to the blue end of the Image, was again more refracted in the second Prism than the Light which went to the red end of that Image, which proves as well the first Proposition as the second. And this happened whether the Axis of the two Prisms were parallel, or inclined to one another, and to the Horizon in any given Angles.

Illustration. Let F [in *Fig.* 18.] be the wide hole in the Window-shut, through which the Sun shines upon the first Prism ABC, and let the refracted Light fall upon the middle of the Board DE, and the middle part of that Light upon the hole G made in the middle part of that Board. Let this trajected part of that Light fall again upon the middle of the second Board *de*, and there paint such an oblong coloured Image of the Sun as was described in the third Experiment. By turning the Prism ABC slowly to and fro about its Axis, this Image will be made to move up and down the Board *de*, and by this means all its parts from one end to the other may be made to pass successively through the hole *g* which is made in the middle of that Board. In the mean while another Prism *abc* is to be fixed next after that hole *g*, to refract the trajected Light a second time. And these things being thus ordered, I marked the places M and N of the opposite Wall upon which the refracted Light fell, and found that whilst the two Boards and second Prism remained unmoved, those places by turning the first Prism about its Axis were changed perpetually. For when the lower part of the Light which fell upon the second Board *de* was cast through

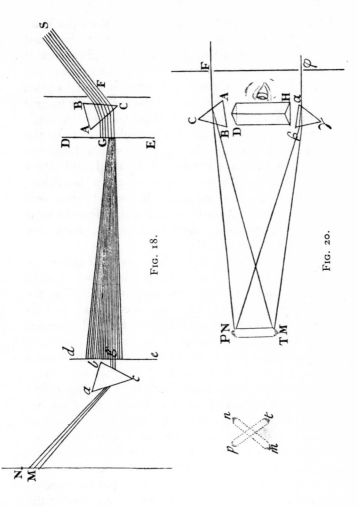

Fig. 18.

Fig. 20.

the hole *g*, it went to a lower place M on the Wall
and when the higher part of that Light was cast
through the same hole *g*, it went to a higher place N
on the Wall, and when any intermediate part of the
Light was cast through that hole, it went to some
place on the Wall between M and N. The unchanged
Position of the holes in the Boards, made the Incidence of the Rays upon the second Prism to be the
same in all cases. And yet in that common Incidence
some of the Rays were more refracted, and others
less. And those were more refracted in this Prism,
which by a greater Refraction in the first Prism were
more turned out of the way, and therefore for their
Constancy of being more refracted are deservedly
called more refrangible.

Exper. 7. At two holes made near one another in
my Window-shut I placed two Prisms, one at each,
which might cast upon the opposite Wall (after the
manner of the third Experiment) two oblong
coloured Images of the Sun. And at a little distance
from the Wall I placed a long slender Paper with
straight and parallel edges, and ordered the Prisms
and Paper so, that the red Colour of one Image
might fall directly upon one half of the Paper, and
the violet Colour of the other Image upon the other
half of the same Paper; so that the Paper appeared of
two Colours, red and violet, much after the manner
of the painted Paper in the first and second Experiments. Then with a black Cloth I covered the Wall
behind the Paper, that no Light might be reflected
from it to disturb the Experiment, and viewing the

Paper through a third Prism held parallel to it, I saw
that half of it which was illuminated by the violet
Light to be divided from the other half by a greater
Refraction, especially when I went a good way off
from the Paper. For when I viewed it too near at
hand, the two halfs of the Paper did not appear fully
divided from one another, but seemed contiguous

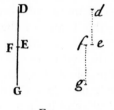

FIG. 19.

at one of their Angles like the painted Paper in the
first Experiment. Which also happened when the
Paper was too broad.

Sometimes instead of the Paper I used a white
Thred, and this appeared through the Prism divided
into two parallel Threds as is represented in the nineteenth Figure, where DG denotes the Thred illuminated with violet Light from D to E and with red
Light from F to G, and *defg* are the parts of the
Thred seen by Refraction. If one half of the Thred
be constantly illuminated with red, and the other
half be illuminated with all the Colours successively,
(which may be done by causing one of the Prisms to
be turned about its Axis whilst the other remains unmoved) this other half in viewing the Thred through

the Prism, will appear in a continual right Line with
the first half when illuminated with red, and begin
to be a little divided from it when illuminated with
Orange, and remove farther from it when illuminated with yellow, and still farther when with green,
and farther when with blue, and go yet farther off
when illuminated with Indigo, and farthest when
with deep violet. Which plainly shews, that the Lights
of several Colours are more and more refrangible one
than another, in this Order of their Colours, red,
orange, yellow, green, blue, indigo, deep violet; and
so proves as well the first Proposition as the second.

I caused also the coloured Spectrums PT [in *Fig.*
17.] and MN made in a dark Chamber by the Refractions of two Prisms to lie in a Right Line end to
end, as was described above in the fifth Experiment,
and viewing them through a third Prism held parallel
to their Length, they appeared no longer in a Right
Line, but became broken from one another, as they
are represented at *pt* and *mn*, the violet end *m* of the
Spectrum *mn* being by a greater Refraction translated farther from its former Place MT than the red
end *t* of the other Spectrum *pt*.

I farther caused those two Spectrums PT [in *Fig.*
20.] and MN to become co-incident in an inverted
Order of their Colours, the red end of each falling
on the violet end of the other, as they are represented
in the oblong Figure PTMN; and then viewing them
through a Prism DH held parallel to their Length,
they appeared not co-incident, as when view'd with
the naked Eye, but in the form of two distinct Spec-

trums *pt* and *mn* crossing one another in the middle
after the manner of the Letter X. Which shews that
the red of the one Spectrum and violet of the other,
which were co-incident at PN and MT, being parted
from one another by a greater Refraction of the
violet to *p* and *m* than of the red to *n* and *t*, do differ
in degrees of Refrangibility.

I illuminated also a little Circular Piece of white
Paper all over with the Lights of both Prisms intermixed, and when it was illuminated with the red of
one Spectrum, and deep violet of the other, so as by
the Mixture of those Colours to appear all over
purple, I viewed the Paper, first at a less distance,
and then at a greater, through a third Prism; and as
I went from the Paper, the refracted Image thereof
became more and more divided by the unequal Refraction of the two mixed Colours, and at length
parted into two distinct Images, a red one and a
violet one, whereof the violet was farthest from the
Paper, and therefore suffered the greatest Refraction. And when that Prism at the Window, which
cast the violet on the Paper was taken away, the
violet Image disappeared; but when the other Prism
was taken away the red vanished; which shews, that
these two Images were nothing else than the Lights
of the two Prisms, which had been intermixed on
the purple Paper, but were parted again by their
unequal Refractions made in the third Prism,
through which the Paper was view'd. This also was
observable, that if one of the Prisms at the Window,
suppose that which cast the violet on the Paper, was

turned about its Axis to make all the Colours in this order, violet, indigo, blue, green, yellow, orange, red, fall successively on the Paper from that Prism, the violet Image changed Colour accordingly, turning successively to indigo, blue, green, yellow and red, and in changing Colour came nearer and nearer to the red Image made by the other Prism, until when it was also red both Images became fully co-incident.

I placed also two Paper Circles very near one another, the one in the red Light of one Prism, and the other in the violet Light of the other. The Circles were each of them an Inch in diameter, and behind them the Wall was dark, that the Experiment might not be disturbed by any Light coming from thence. These Circles thus illuminated, I viewed through a Prism, so held, that the Refraction might be made towards the red Circle, and as I went from them they came nearer and nearer together, and at length became co-incident; and afterwards when I went still farther off, they parted again in a contrary Order, the violet by a greater Refraction being carried beyond the red.

Exper. 8. In Summer, when the Sun's Light uses to be strongest, I placed a Prism at the Hole of the Window-shut, as in the third Experiment, yet so that its Axis might be parallel to the Axis of the World, and at the opposite Wall in the Sun's refracted Light, I placed an open Book. Then going six Feet and two Inches from the Book, I placed there the above-mentioned Lens, by which the Light reflected from the Book might be made to converge

and meet again at the distance of six Feet and two
Inches behind the Lens, and there paint the Species
of the Book upon a Sheet of white Paper much after
the manner of the second Experiment. The Book and
Lens being made fast, I noted the Place where the
Paper was, when the Letters of the Book, illuminated
by the fullest red Light of the Solar Image falling
upon it, did cast their Species on that Paper most
distinctly: And then I stay'd till by the Motion of the
Sun, and consequent Motion of his Image on the
Book, all the Colours from that red to the middle of
the blue pass'd over those Letters; and when those
Letters were illuminated by that blue, I noted again
the Place of the Paper when they cast their Species
most distinctly upon it: And I found that this last
Place of the Paper was nearer to the Lens than its
former Place by about two Inches and an half, or two
and three quarters. So much sooner therefore did the
Light in the violet end of the Image by a greater Refraction converge and meet, than the Light in the
red end. But in trying this, the Chamber was as dark
as I could make it. For, if these Colours be diluted
and weakned by the Mixture of any adventitious
Light, the distance between the Places of the Paper
will not be so great. This distance in the second Experiment, where the Colours of natural Bodies were
made use of, was but an Inch and an half, by reason
of the Imperfection of those Colours. Here in the
Colours of the Prism, which are manifestly more full,
intense, and lively than those of natural Bodies, the
distance is two Inches and three quarters. And were

the Colours still more full, I question not but that the distance would be considerably greater. For the coloured Light of the Prism, by the interfering of the Circles described in the second Figure of the fifth Experiment, and also by the Light of the very bright Clouds next the Sun's Body intermixing with these Colours, and by the Light scattered by the In-equalities in the Polish of the Prism, was so very much compounded, that the Species which those faint and dark Colours, the indigo and violet, cast upon the Paper were not distinct enough to be well observed.

Exper. 9. A Prism, whose two Angles at its Base were equal to one another, and half right ones, and the third a right one, I placed in a Beam of the Sun's Light let into a dark Chamber through a Hole in the Window-shut, as in the third Experiment. And turning the Prism slowly about its Axis, until all the Light which went through one of its Angles, and was refracted by it began to be reflected by its Base, at which till then it went out of the Glass, I observed that those Rays which had suffered the greatest Re-fraction were sooner reflected than the rest. I conceived therefore, that those Rays of the reflected Light, which were most refrangible, did first of all by a total Reflexion become more copious in that Light than the rest, and that afterwards the rest also, by a total Reflexion, became as copious as these. To try this, I made the reflected Light pass through another Prism, and being refracted by it to fall afterwards upon a Sheet of white Paper placed

at some distance behind it, and there by that
Refraction to paint the usual Colours of the Prism.
And then causing the first Prism to be turned about
its Axis as above, I observed that when those Rays,
which in this Prism had suffered the greatest
Refraction, and appeared of a blue and violet Colour
began to be totally reflected, the blue and violet
Light on the Paper, which was most refracted in the
second Prism, received a sensible Increase above
that of the red and yellow, which was least refracted;
and afterwards, when the rest of the Light which
was green, yellow, and red, began to be totally reflected in the first Prism, the Light of those Colours
on the Paper received as great an Increase as the
violet and blue had done before. Whence 'tis manifest, that the Beam of Light reflected by the Base of
the Prism, being augmented first by the more refrangible Rays, and afterwards by the less refrangible ones, is compounded of Rays differently refrangible. And that all such reflected Light is of the
same Nature with the Sun's Light before its Incidence on the Base of the Prism, no Man ever doubted;
it being generally allowed, that Light by such
Reflexions suffers no Alteration in its Modifications
and Properties. I do not here take Notice of any Refractions made in the sides of the first Prism, because
the Light enters it perpendicularly at the first side,
and goes out perpendicularly at the second side, and
therefore suffers none. So then, the Sun's incident
Light being of the same Temper and Constitution
with his emergent Light, and the last being com-

pounded of Rays differently refrangible, the first
must be in like manner compounded.

Illustration. In the twenty-first Figure, ABC is the
first Prism, BC its Base, B and C its equal Angles at
the Base, each of 45 Degrees, A its rectangular
Vertex, FM a beam of the Sun's Light let into a
dark Room through a hole F one third part of an

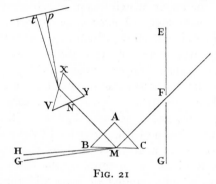

FIG. 21

Inch broad, M its Incidence on the Base of the
Prism, MG a less refracted Ray, MH a more refracted Ray, MN the beam of Light reflected from
the Base, VXY the second Prism by which this beam
in passing through it is refracted, N*t* the less refracted Light of this beam, and N*p* the more refracted
part thereof. When the first Prism ABC is turned
about its Axis according to the order of the Letters
ABC, the Rays MH emerge more and more obliquely out of that Prism, and at length after their
most oblique Emergence are reflected towards N,
and going on to *p* do increase the Number of the

Rays N*p*. Afterwards by continuing the Motion of
the first Prism, the Rays MG are also reflected to N
and increase the number of the Rays N*t*. And therefore the Light MN admits into its Composition, first
the more refrangible Rays, and then the less refrangible Rays, and yet after this Composition is of
the same Nature with the Sun's immediate Light
FM, the Reflexion of the specular Base BC causing
no Alteration therein.

Exper. 10. Two Prisms, which were alike in Shape,
I tied so together, that their Axis and opposite Sides
being parallel, they composed a Parallelopiped. And,
the Sun shining into my dark Chamber through a
little hole in the Window-shut, I placed that Parallelopiped in his beam at some distance from the hole,
in such a Posture, that the Axes of the Prisms might
be perpendicular to the incident Rays, and that those
Rays being incident upon the first Side of one Prism,
might go on through the two contiguous Sides of
both Prisms, and emerge out of the last Side of the
second Prism. This Side being parallel to the first
Side of the first Prism, caused the emerging Light
to be parallel to the incident. Then, beyond these
two Prisms I placed a third, which might refract that
emergent Light, and by that Refraction cast the
usual Colours of the Prism upon the opposite Wall,
or upon a sheet of white Paper held at a convenient
Distance behind the Prism for that refracted Light
to fall upon it. After this I turned the Parallelopiped
about its Axis, and found that when the contiguous
Sides of the two Prisms became so oblique to the

incident Rays, that those Rays began all of them to be reflected, those Rays which in the third Prism had suffered the greatest Refraction, and painted the Paper with violet and blue, were first of all by a total Reflexion taken out of the transmitted Light, the rest remaining and on the Paper painting their Colours of green, yellow, orange and red, as before; and afterwards by continuing the Motion of the two Prisms, the rest of the Rays also by a total Reflexion vanished in order, according to their degrees of Refrangibility. The Light therefore which emerged out of the two Prisms is compounded of Rays differently refrangible, seeing the more refrangible Rays may be taken out of it, while the less refrangible remain. But this Light being trajected only through the parallel Superficies of the two Prisms, if it suffer'd any change by the Refraction of one Superficies it lost that Impression by the contrary Refraction of the other Superficies, and so being restor'd to its pristine Constitution, became of the same Nature and Condition as at first before its Incidence on those Prisms; and therefore, before its Incidence, was as much compounded of Rays differently refrangible, as afterwards.

Illustration. In the twenty second Figure ABC and BCD are the two Prisms tied together in the form of a Parallelopiped, their Sides BC and CB being contiguous, and their Sides AB and CD parallel. And HJK is the third Prism, by which the Sun's Light propagated through the hole F into the dark Chamber, and there passing through those sides

of the Prisms AB, BC, CB and CD, is refracted at O
to the white Paper PT, falling there partly upon P
by a greater Refraction, partly upon T by a less Refraction, and partly upon R and other intermediate
places by intermediate Refractions. By turning the

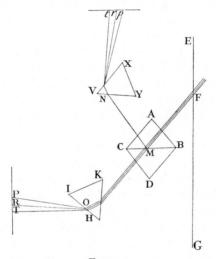

FIG. 22.

Parallelopiped ACBD about its Axis, according to
the order of the Letters A, C, D, B, at length when
the contiguous Planes BC and CB become sufficiently oblique to the Rays FM, which are incident
upon them at M, there will vanish totally out of the
refracted Light OPT, first of all the most refracted
Rays OP, (the rest OR and OT remaining as before)
then the Rays OR and other intermediate ones, and

lastly, the least refracted Rays OT. For when the Plane BC becomes sufficiently oblique to the Rays incident upon it, those Rays will begin to be totally reflected by it towards N; and first the most refrangible Rays will be totally reflected (as was explained in the preceding Experiment) and by Consequence must first disappear at P, and afterwards the rest as they are in order totally reflected to N, they must disappear in the same order at R and T. So then the Rays which at O suffer the greatest Refraction, may be taken out of the Light MO whilst the rest of the Rays remain in it, and therefore that Light MO is compounded of Rays differently refrangible. And because the Planes AB and CD are parallel, and therefore by equal and contrary Refractions destroy one anothers Effects, the incident Light FM must be of the same Kind and Nature with the emergent Light MO, and therefore doth also consist of Rays differently refrangible. These two Lights FM and MO, before the most refrangible Rays are separated out of the emergent Light MO, agree in Colour, and in all other Properties so far as my Observation reaches, and therefore are deservedly reputed of the same Nature and Constitution, and by Consequence the one is compounded as well as the other. But after the most refrangible Rays begin to be totally reflected, and thereby separated out of the emergent Light MO, that Light changes its Colour from white to a dilute and faint yellow, a pretty good orange, a very full red successively, and then totally vanishes. For after the most refrangible Rays which paint the

Paper at P with a purple Colour, are by a total Reflexion taken out of the beam of Light MO, the rest
of the Colours which appear on the Paper at R and
T being mix'd in the Light MO compound there a
faint yellow, and after the blue and part of the green
which appear on the Paper between P and R are
taken away, the rest which appear between R and T
(that is the yellow, orange, red and a little green)
being mixed in the beam MO compound there an
orange; and when all the Rays are by Reflexion taken
out of the beam MO, except the least refrangible,
which at T appear of a full red, their Colour is the
same in that beam MO as afterwards at T, the Refraction of the Prism HJK serving only to separate
the differently refrangible Rays, without making
any Alteration in their Colours, as shall be more
fully proved hereafter. All which confirms as well the
first Proposition as the second.

Scholium. If this Experiment and the former be
conjoined and made one by applying a fourth Prism
VXY [in *Fig.* 22.] to refract the reflected beam MN
towards *tp*, the Conclusion will be clearer. For then
the Light N*p* which in the fourth Prism is more refracted, will become fuller and stronger when the
Light OP, which in the third Prism HJK is more refracted, vanishes at P; and afterwards when the less
refracted Light OT vanishes at T, the less refracted
Light N*t* will become increased whilst the more refracted Light at *p* receives no farther increase. And as
the trajected beam MO in vanishing is always of such
a Colour as ought to result from the mixture of the

Colours which fall upon the Paper PT, so is the reflected beam MN always of such a Colour as ought to result from the mixture of the Colours which fall upon the Paper *pt*. For when the most refrangible Rays are by a total Reflexion taken out of the beam MO, and leave that beam of an orange Colour, the Excess of those Rays in the reflected Light, does not only make the violet, indigo and blue at *p* more full, but also makes the beam MN change from the yellowish Colour of the Sun's Light, to a pale white inclining to blue, and afterward recover its yellowish Colour again, so soon as all the rest of the transmitted Light MOT is reflected.

Now seeing that in all this variety of Experiments, whether the Trial be made in Light reflected, and that either from natural Bodies, as in the first and second Experiment, or specular, as in the ninth; or in Light refracted, and that either before the unequally refracted Rays are by diverging separated from one another, and losing their whiteness which they have altogether, appear severally of several Colours, as in the fifth Experiment; or after they are separated from one another, and appear colour'd as in the sixth, seventh, and eighth Experiments; or in Light trajected through parallel Superficies, destroying each others Effects, as in the tenth Experiment; there are always found Rays, which at equal Incidences on the same Medium suffer unequal Refractions, and that without any splitting or dilating of single Rays, or contingence in the inequality of the Refractions, as is proved in the fifth and sixth

Experiments. And seeing the Rays which differ in Refrangibility may be parted and sorted from one another, and that either by Refraction as in the third Experiment, or by Reflexion as in the tenth, and then the several sorts apart at equal Incidences suffer unequal Refractions, and those sorts are more refracted than others after Separation, which were more refracted before it, as in the sixth and following Experiments, and if the Sun's Light be trajected through three or more cross Prisms successively, those Rays which in the first Prism are refracted more than others, are in all the following Prisms refracted more than others in the same Rate and Proportion, as appears by the fifth Experiment; it's manifest that the Sun's Light is an heterogeneous Mixture of Rays, some of which are constantly more refrangible than others, as was proposed.

PROP. III. Theor. III.

The Sun's Light consists of Rays differing in Reflexibility, and those Rays are more reflexible than others which are more refrangible.

THIS is manifest by the ninth and tenth Experiments: For in the ninth Experiment, by turning the Prism about its Axis, until the Rays within it which in going out into the Air were refracted by its Base, became so oblique to that Base, as to begin to be totally reflected thereby; those Rays became

first of all totally reflected, which before at equal
Incidences with the rest had suffered the greatest
Refraction. And the same thing happens in the
Reflexion made by the common Base of the two
Prisms in the tenth Experiment.

PROP. IV. PROB. I.

*To separate from one another the heterogeneous Rays
of compound Light.*

THE heterogeneous Rays are in some measure
separated from one another by the Refraction
of the Prism in the third Experiment, and in the fifth
Experiment, by taking away the Penumbra from the
rectilinear sides of the coloured Image, that Separation in those very rectilinear sides or straight edges
of the Image becomes perfect. But in all places between those rectilinear edges, those innumerable
Circles there described, which are severally illuminated by homogeneal Rays, by interfering with one
another, and being every where commix'd, do render
the Light sufficiently compound. But if these Circles,
whilst their Centers keep their Distances and Positions, could be made less in Diameter, their interfering one with another, and by Consequence the
Mixture of the heterogeneous Rays would be proportionally diminish'd. In the twenty third Figure let
AG, BH, CJ, DK, EL, FM be the Circles which so
many sorts of Rays flowing from the same disque of

the Sun, do in the third Experiment illuminate; of all which and innumerable other intermediate ones lying in a continual Series between the two rectilinear and parallel edges of the Sun's oblong Image PT, that Image is compos'd, as was explained in the fifth

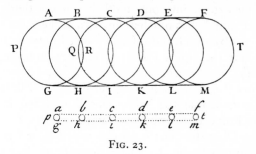

FIG. 23.

Experiment. And let *ag*, *bh*, *ci*, *dk*, *el*, *fm* be so many less Circles lying in a like continual Series between two parallel right Lines *af* and *gm* with the same distances between their Centers, and illuminated by the same sorts of Rays, that is the Circle *ag* with the same sort by which the corresponding Circle AG was illuminated, and the Circle *bh* with the same sort by which the corresponding Circle BH was illuminated, and the rest of the Circles *ci*, *dk*, *el*, *fm* respectively, with the same sorts of Rays by which the several corresponding Circles CJ, DK, EL, FM were illuminated. In the Figure PT composed of the greater Circles, three of those Circles AG, BH, CJ, are so expanded into one another, that the three sorts of Rays by which those Circles are illuminated, together with other innumerable sorts of intermediate

Rays, are mixed at QR in the middle of the Circle
BH. And the like Mixture happens throughout
almost the whole length of the Figure PT. But in the
Figure *pt* composed of the less Circles, the three less
Circles *ag*, *bh*, *ci*, which answer to those three greater,
do not extend into one another; nor are there any
where mingled so much as any two of the three sorts
of Rays by which those Circles are illuminated, and
which in the Figure PT are all of them intermingled
at BH.

Now he that shall thus consider it, will easily
understand that the Mixture is diminished in the
same Proportion with the Diameters of the Circles.
If the Diameters of the Circles whilst their Centers
remain the same, be made three times less than
before, the Mixture will be also three times less;
if ten times less, the Mixture will be ten times less,
and so of other Proportions. That is, the Mixture of
the Rays in the greater Figure PT will be to their
Mixture in the less *pt*, as the Latitude of the greater
Figure is to the Latitude of the less. For the Lati-
tudes of these Figures are equal to the Diameters of
their Circles. And hence it easily follows, that the
Mixture of the Rays in the refracted Spectrum *pt*
is to the Mixture of the Rays in the direct and immediate Light of the Sun, as the breadth of that
Spectrum is to the difference between the length and
breadth of the same Spectrum.

So then, if we would diminish the Mixture of the
Rays, we are to diminish the Diameters of the Circles.
Now these would be diminished if the Sun's Dia-

NEWTON WHILE BACHELOR OF ARTS IN
TRINITY COLLEGE, CAMBRIDGE

NEWTON IN 1706, TWO YEARS
AFTER THE PUBLICATION OF "OPTICKS"

meter to which they answer could be made less than
it is, or (which comes to the same Purpose) if without
Doors, at a great distance from the Prism towards
the Sun, some opake Body were placed, with a round
hole in the middle of it, to intercept all the Sun's
Light, excepting so much as coming from the middle
of his Body could pass through that Hole to the
Prism. For so the Circles AG, BH, and the rest,
would not any longer answer to the whole Disque
of the Sun, but only to that Part of it which could be
seen from the Prism through that Hole, that it is to
the apparent Magnitude of that Hole view'd from
the Prism. But that these Circles may answer more
distinctly to that Hole, a Lens is to be placed by the
Prism to cast the Image of the Hole, (that is, every
one of the Circles AG, BH, &c.) distinctly upon the
Paper at PT, after such a manner, as by a Lens placed
at a Window, the Species of Objects abroad are cast
distinctly upon a Paper within the Room, and the
rectilinear Sides of the oblong Solar Image in the
fifth Experiment became distinct without any Penumbra. If this be done, it will not be necessary to
place that Hole very far off, no not beyond the Window. And therefore instead of that Hole, I used the
Hole in the Window-shut, as follows.

Exper. 11. In the Sun's Light let into my darken'd
Chamber through a small round Hole in my Window-
shut, at about ten or twelve Feet from the Window,
I placed a Lens, by which the Image of the Hole
might be distinctly cast upon a Sheet of white Paper,
placed at the distance of six, eight, ten, or twelve

Feet from the Lens. For, according to the difference
of the Lenses I used various distances, which I think
not worth the while to describe. Then immediately
after the Lens I placed a Prism, by which the trajected Light might be refracted either upwards or
side-ways, and thereby the round Image, which the
Lens alone did cast upon the Paper might be drawn
out into a long one with Parallel Sides, as in the third
Experiment. This oblong Image I let fall upon
another Paper at about the same distance from the
Prism as before, moving the Paper either towards the
Prism or from it, until I found the just distance
where the Rectilinear Sides of the Image became
most distinct. For in this Case, the Circular Images
of the Hole, which compose that Image after the
same manner that the Circles *ag*, *bh*, *ci*, &c. do the
Figure *pt* [in *Fig.* 23.] were terminated most distinctly without any Penumbra, and therefore extended into one another the least that they could,
and by consequence the Mixture of the heterogeneous Rays was now the least of all. By this means I
used to form an oblong Image (such as is *pt*) [in *Fig.*
23, and 24.] of Circular Images of the Hole, (such as
are *ag*, *bh*, *ci*, &c.) and by using a greater or less Hole
in the Window-shut, I made the Circular Images *ag*,
bh, *ci*, &c. of which it was formed, to become greater
or less at pleasure, and thereby the Mixture of the
Rays in the Image *pt* to be as much, or as little as I
desired.

Illustration. In the twenty-fourth Figure, F represents the Circular Hole in the Window-shut, MN

the Lens, whereby the Image or Species of that Hole
is cast distinctly upon a Paper at J, ABC the Prism,
whereby the Rays are at their emerging out of the
Lens refracted from J towards another Paper at *pt*,
and the round Image at J is turned into an oblong

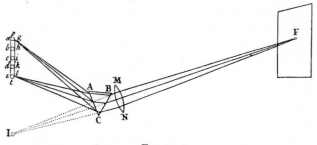

FIG. 24.

Image *pt* falling on that other Paper. This Image *pt*
consists of Circles placed one after another in a
Rectilinear Order, as was sufficiently explained in
the fifth Experiment; and these Circles are equal to
the Circle J, and consequently answer in magnitude
to the Hole F; and therefore by diminishing that
Hole they may be at pleasure diminished, whilst their
Centers remain in their Places. By this means I made
the Breadth of the Image *pt* to be forty times, and
sometimes sixty or seventy times less than its Length.
As for instance, if the Breadth of the Hole F be one
tenth of an Inch, and MF the distance of the Lens
from the Hole be 12 Feet; and if *p*B or *p*M the distance of the Image *pt* from the Prism or Lens be 10
Feet, and the refracting Angle of the Prism be 62

Degrees, the Breadth of the Image *pt* will be one twelfth of an Inch, and the Length about six Inches, and therefore the Length to the Breadth as 72 to 1, and by consequence the Light of this Image 71 times less compound than the Sun's direct Light. And Light thus far simple and homogeneal, is sufficient for trying all the Experiments in this Book about simple Light. For the Composition of heterogeneal Rays is in this Light so little, that it is scarce to be discovered and perceiv'd by Sense, except perhaps in the indigo and violet. For these being dark Colours do easily suffer a sensible Allay by that little scattering Light which uses to be refracted irregularly by the Inequalities of the Prism.

Yet instead of the Circular Hole F, 'tis better to substitute an oblong Hole shaped like a long Parallelogram with its Length parallel to the Prism ABC. For if this Hole be an Inch or two long, and but a tenth or twentieth Part of an Inch broad, or narrower; the Light of the Image *pt* will be as simple as before, or simpler, and the Image will become much broader, and therefore more fit to have Experiments try'd in its Light than before.

Instead of this Parallelogram Hole may be substituted a triangular one of equal Sides, whose Base, for instance, is about the tenth Part of an Inch, and its Height an Inch or more. For by this means, if the Axis of the Prism be parallel to the Perpendicular of the Triangle, the Image *pt* [in *Fig.* 25.] will now be form'd of equicrural Triangles *ag*, *bh*, *ci*, *dk*, *el*, *fm*, &c. and innumerable other intermediate ones answering to the triangular Hole in Shape and Bigness, and lying one after another in a continual Series between two Parallel Lines *af* and *gm*. These Triangles are a little intermingled at their Bases, but not at their Vertices; and therefore the Light on the brighter

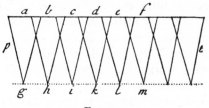

FIG. 25.

Side *af* of the Image, where the Bases of the Triangles are, is a little compounded, but on the darker Side *gm* is altogether uncompounded, and in all Places between the Sides the Composition is proportional to the distances of the Places from that obscurer Side *gm*. And having a Spectrum *pt* of such a Composition, we may try Experiments either in its stronger and less simple Light near the Side *af*, or in its weaker and simpler Light near the other Side *gm*, as it shall seem most convenient.

But in making Experiments of this kind, the Chamber ought to be made as dark as can be, lest any Foreign Light mingle it self with the Light of the Spectrum *pt*, and render it compound; especially if we would try Experiments in the more simple Light next the Side *gm* of the Spectrum; which being fainter, will have a less proportion to the Foreign Light; and so by the mixture of that Light be more troubled, and made more compound. The Lens also ought to be good, such as may serve for optical Uses, and the Prism ought to have a large Angle, suppose of 65 or 70 Degrees, and to be well wrought, being made of Glass free from Bubbles and Veins, with its Sides not a little convex or concave, as usually happens, but truly plane, and its Polish elaborate, as in working Optick-glasses, and not such as is usually wrought with Putty, whereby the edges of the Sand-holes being worn away, there are left all over the Glass a numberless Company of very little convex polite Risings like Waves. The edges also of the Prism and Lens, so far as they may make any irregular Refraction, must be covered with a black Paper glewed on. And all the Light of the Sun's Beam let into the Chamber, which is useless and unprofitable to the Experiment, ought to be intercepted with black Paper, or other black Obstacles. For otherwise the useless Light being reflected every way in the Chamber, will mix with the oblong Spectrum, and help to disturb it. In trying these Things, so much diligence is not altogether necessary, but it will promote the Success of the Experiments, and by a very scrupulous Examiner of Things deserves to be apply'd. It's difficult to get Glass Prisms fit for this Purpose, and therefore I used sometimes prismatick Vessels made with pieces of broken Looking-glasses, and filled with Rain Water. And to increase the Refraction, I sometimes impregnated the Water strongly with *Saccharum Saturni*.

PROP. V. Theor. IV.

Homogeneal Light is refracted regularly without any
Dilatation splitting or shattering of the Rays, and
the confused Vision of Objects seen through re-
fracting Bodies by heterogeneal Light arises from
the different Refrangibility of several sorts of Rays.

THE first Part of this Proposition has been already sufficiently proved in the fifth Experiment, and will farther appear by the Experiments
which follow.

Exper. 12. In the middle of a black Paper I made
a round Hole about a fifth or sixth Part of an Inch
in diameter. Upon this Paper I caused the Spectrum
of homogeneal Light described in the former Proposition, so to fall, that some part of the Light might
pass through the Hole of the Paper. This transmitted
part of the Light I refracted with a Prism placed
behind the Paper, and letting this refracted Light
fall perpendicularly upon a white Paper two or three
Feet distant from the Prism, I found that the Spectrum formed on the Paper by this Light was not
oblong, as when 'tis made (in the third Experiment)
by refracting the Sun's compound Light, but was
(so far as I could judge by my Eye) perfectly circular,
the Length being no greater than the Breadth.
Which shews, that this Light is refracted regularly
without any Dilatation of the Rays.

Exper. 13. In the homogeneal Light I placed a
Paper Circle of a quarter of an Inch in diameter, and

in the Sun's unrefracted heterogeneal white Light
I placed another Paper Circle of the same Bigness.
And going from the Papers to the distance of some
Feet, I viewed both Circles through a Prism. The
Circle illuminated by the Sun's heterogeneal Light
appeared very oblong, as in the fourth Experiment,
the Length being many times greater than the
Breadth; but the other Circle, illuminated with homogeneal Light, appeared circular and distinctly defined, as when 'tis view'd with the naked Eye. Which
proves the whole Proposition.

Exper. 14. In the homogeneal Light I placed Flies,
and such-like minute Objects, and viewing them
through a Prism, I saw their Parts as distinctly defined, as if I had viewed them with the naked Eye.
The same Objects placed in the Sun's unrefracted
heterogeneal Light, which was white, I viewed also
through a Prism, and saw them most confusedly defined, so that I could not distinguish their smaller
Parts from one another. I placed also the Letters of
a small print, one while in the homogeneal Light,
and then in the heterogeneal, and viewing them
through a Prism, they appeared in the latter Case so
confused and indistinct, that I could not read them;
but in the former they appeared so distinct, that I
could read readily, and thought I saw them as distinct, as when I view'd them with my naked Eye.
In both Cases I view'd the same Objects, through
the same Prism at the same distance from me, and
in the same Situation. There was no difference, but
in the Light by which the Objects were illuminated,

and which in one Case was simple, and in the other compound; and therefore, the distinct Vision in the former Case, and confused in the latter, could arise from nothing else than from that difference of the Lights. Which proves the whole Proposition.

And in these three Experiments it is farther very remarkable, that the Colour of homogeneal Light was never changed by the Refraction.

PROP. VI. THEOR. V.

The Sine of Incidence of every Ray considered apart, is to its Sine of Refraction in a given Ratio.

THAT every Ray consider'd apart, is constant to it self in some degree of Refrangibility, is sufficiently manifest out of what has been said. Those Rays, which in the first Refraction, are at equal Incidences most refracted, are also in the following Refractions at equal Incidences most refracted; and so of the least refrangible, and the rest which have any mean Degree of Refrangibility, as is manifest by the fifth, sixth, seventh, eighth, and ninth Experiments. And those which the first Time at like Incidences are equally refracted, are again at like Incidences equally and uniformly refracted, and that whether they be refracted before they be separated from one another, as in the fifth Experiment, or whether they be refracted apart, as in the twelfth, thirteenth and fourteenth Experiments. The Refraction therefore of every Ray apart is regular, and what Rule that Refraction observes we are now to shew.*

The late Writers in Opticks teach, that the Sines of Incidence are in a given Proportion to the Sines of Refraction, as was explained in the fifth Axiom, and some by Instruments fitted for measuring of Refractions, or otherwise experimentally examining this Proportion, do acquaint us that they have found it accurate. But whilst they, not understanding the different Refrangibility of several Rays, conceived them all to be refracted according to one and the same Proportion, 'tis to be presumed that they adapted their Measures only to the middle of the refracted Light; so that from their Measures we may conclude only that the Rays which have a mean Degree of Refrangibility, that is, those which when separated from the rest appear green, are refracted according to a given Proportion of their Sines. And therefore we are now to shew, that the like given Proportions obtain in all the rest. That it should be so is very reasonable, Nature being ever conformable to her self; but an experimental Proof is desired. And such a Proof will be had, if we can shew that the Sines of Refraction of Rays differently refrangible are one to another in a given Proportion when their Sines of Incidence are equal. For, if the Sines of Refraction of all the Rays are in given Proportions to the Sine of Refractions of a Ray which has a mean Degree of Refrangibility, and this Sine is in a given

* *This is very fully treated of in our* Author's Lect. Optic. *Part* I. *Sect.* II.

Proportion to the equal Sines of Incidence, those
other Sines of Refraction will also be in given Proportions to the equal Sines of Incidence. Now, when
the Sines of Incidence are equal, it will appear by
the following Experiment, that the Sines of Refraction are in a given Proportion to one another.

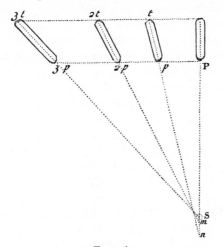

Fig. 26.

Exper. 15. The Sun shining into a dark Chamber
through a little round Hole in the Window-shut, let
S [in *Fig.* 26.] represent his round white Image
painted on the opposite Wall by his direct Light,
PT his oblong coloured Image made by refracting
that Light with a Prism placed at the Window; and
pt, or 2*p* 2*t*, 3*p* 3*t*, his oblong colour'd Image made
by refracting again the same Light side-ways with a

second Prism placed immediately after the first in a cross Position to it, as was explained in the fifth Experiment; that is to say, *pt* when the Refraction of the second Prism is small, *2p 2t* when its Refraction is greater, and *3p 3t* when it is greatest. For such will be the diversity of the Refractions, if the refracting Angle of the second Prism be of various Magnitudes; suppose of fifteen or twenty Degrees to make the Image *pt*, of thirty or forty to make the Image *2p 2t*, and of sixty to make the Image *3p 3t*. But for want of solid Glass Prisms with Angles of convenient Bignesses, there may be Vessels made of polished Plates of Glass cemented together in the form of Prisms and filled with Water. These things being thus ordered, I observed that all the solar Images or coloured Spectrums PT, *pt*, *2p 2t*, *3p 3t* did very nearly converge to the place S on which the direct Light of the Sun fell and painted his white round Image when the Prisms were taken away. The Axis of the Spectrum PT, that is the Line drawn through the middle of it parallel to its rectilinear Sides, did when produced pass exactly through the middle of that white round Image S. And when the Refraction of the second Prism was equal to the Refraction of the first, the refracting Angles of them both being about 60 Degrees, the Axis of the Spectrum *3p 3t* made by that Refraction, did when produced pass also through the middle of the same white round Image S. But when the Refraction of the second Prism was less than that of the first, the produced Axes of the Spectrums *tp* or *2t 2p* made by that Re-

fraction did cut the produced Axis of the Spectrum
TP in the points *m* and *n*, a little beyond the Center
of that white round Image S. Whence the proportion
of the Line $3tT$ to the Line $3pP$ was a little greater
than the Proportion of $2tT$ or $2pP$, and this Proportion a little greater than that of tT to pP. Now when
the Light of the Spectrum PT falls perpendicularly
upon the Wall, those Lines $3tT$, $3pP$, and $2tT$, $2pP$,
and tT, pP, are the Tangents of the Refractions, and
therefore by this Experiment the Proportions of the
Tangents of the Refractions are obtained, from
whence the Proportions of the Sines being derived,
they come out equal, so far as by viewing the Spectrums, and using some mathematical Reasoning I
could estimate. For I did not make an accurate Computation. So then the Proposition holds true in every
Ray apart, so far as appears by Experiment. And
that it is accurately true, may be demonstrated upon
this Supposition. *That Bodies refract Light by acting
upon its Rays in Lines perpendicular to their Surfaces.*
But in order to this Demonstration, I must distinguish the Motion of every Ray into two Motions, the
one perpendicular to the refracting Surface, the
other parallel to it, and concerning the perpendicular
Motion lay down the following Proposition.

   If any Motion or moving thing whatsoever be incident with any Velocity on any broad and thin
space terminated on both sides by two parallel Planes,
and in its Passage through that space be urged perpendicularly towards the farther Plane by any force
which at given distances from the Plane is of given

Quantities; the perpendicular velocity of that Motion or Thing, at its emerging out of that space, shall be always equal to the square Root of the sum of the square of the perpendicular velocity of that Motion or Thing at its Incidence on that space; and of the square of the perpendicular velocity which that Motion or Thing would have at its Emergence, if at its Incidence its perpendicular velocity was infinitely little.

And the same Proposition holds true of any Motion or Thing perpendicularly retarded in its passage through that space, if instead of the sum of the two Squares you take their difference. The Demonstration Mathematicians will easily find out, and therefore I shall not trouble the Reader with it.

Suppose now that a Ray coming most obliquely in the Line MC [in *Fig*. 1.] be refracted at C by the Plane RS into the Line CN, and if it be required to find the Line CE, into which any other Ray AC shall be refracted; let MC, AD, be the Sines of Incidence of the two Rays, and NG, EF, their Sines of Refraction, and let the equal Motions of the incident Rays be represented by the equal Lines MC and AC, and the Motion MC being considered as parallel to the refracting Plane, let the other Motion AC be distinguished into two Motions AD and DC, one of which AD is parallel, and the other DC perpendicular to the refracting Surface. In like manner, let the Motions of the emerging Rays be distinguish'd into two, whereof the perpendicular ones are $\frac{MC}{NG}$ CG and $\dfrac{AD}{EF}$ CF. And if the force of the refracting Plane begins to act upon the Rays either in that Plane or at a certain distance from it on the one side, and ends at a certain distance from it on the other side, and in all places between those two limits acts upon the Rays in Lines perpendicular to that refracting Plane, and the Actions upon the Rays at equal distances from the refracting Plane be equal, and at unequal ones either equal or unequal according to any rate whatever; that Motion of the Ray which is parallel to the refracting Plane, will suffer no Alteration by that Force; and that Motion which is perpendicular to it will be altered according to the rule of the fore-going Proposition. If therefore for the perpendicular velocity of the emerging Ray CN you write $\dfrac{MC}{NG}$ CG as above, then the perpendicular velocity of any other emerging Ray CE which was $\dfrac{AD}{EF}$ CF, will be equal to the square Root of $CDq + \dfrac{MCq}{NGq} CGq$. And by squaring these Equals, and adding to them the Equals ADq and $MCq - CDq$, and dividing the Sums by the Equals $CFq + EFq$ and $CGq + NGq$, you will have $\dfrac{MCq}{NGq}$ equal to $\dfrac{ADq}{EFq}$. Whence AD, the Sine of Incidence, is to EF the Sine of Refraction, as MC to NG, that is, in a given *ratio*. And this Demonstration being general, without determining what Light

N.O.

is, or by what kind of Force it is refracted, or assuming any thing farther than that the refracting Body acts upon the Rays in Lines perpendicular to its Surface; I take it to be a very convincing Argument of the full truth of this Proposition.

So then, if the *ratio* of the Sines of Incidence and Refraction of any sort of Rays be found in any one case, 'tis given in all cases; and this may be readily found by the Method in the following Proposition.

PROP. VII. THEOR. VI.

The Perfection of Telescopes is impeded by the different Refrangibility of the Rays of Light.

THE Imperfection of Telescopes is vulgarly attributed to the spherical Figures of the Glasses, and therefore Mathematicians have propounded to figure them by the conical Sections. To shew that they are mistaken, I have inserted this Proposition; the truth of which will appear by the measure of the Refractions of the several sorts of Rays; and these measures I thus determine.

In the third Experiment of this first Part, where the refracting Angle of the Prism was 62½ Degrees, the half of that Angle 31 deg. 15 min. is the Angle of Incidence of the Rays at their going out of the Glass into the Air*; and the Sine of this Angle is 5188, the Radius being 10000. When the Axis of this

---

* *See our* Author's Lect. Optic. Part I. Sect. II. § 29.

Prism was parallel to the Horizon, and the Refraction of the Rays at their Incidence on this Prism equal to that at their Emergence out of it, I observed with a Quadrant the Angle which the mean refrangible Rays, (that is those which went to the middle of the Sun's coloured Image) made with the Horizon, and by this Angle and the Sun's altitude observed at the same time, I found the Angle which the emergent Rays contained with the incident to be 44 deg. and 40 min. and the half of this Angle added to the Angle of Incidence 31 deg. 15 min. makes the Angle of Refraction, which is therefore 53 deg. 35 min. and its Sine 8047. These are the Sines of Incidence and Refraction of the mean refrangible Rays, and their Proportion in round Numbers is 20 to 31. This Glass was of a Colour inclining to green. The last of the Prisms mentioned in the third Experiment was of clear white Glass. Its refracting Angle 63½ Degrees. The Angle which the emergent Rays contained, with the incident 45 deg. 50 min. The Sine of half the first Angle 5262. The Sine of half the Sum of the Angles 8157. And their Proportion in round Numbers 20 to 31, as before.

From the Length of the Image, which was about 9¾ or 10 Inches, subduct its Breadth, which was 2⅛ Inches, and the Remainder 7¾ Inches would be the Length of the Image were the Sun but a Point, and therefore subtends the Angle which the most and least refrangible Rays, when incident on the Prism in the same Lines, do contain with one another after their Emergence. Whence this Angle is 2 deg. 0′. 7″.

For the distance between the Image and the Prism where this Angle is made, was $18\frac{1}{2}$ Feet, and at that distance the Chord $7\frac{3}{4}$ Inches subtends an Angle of 2 deg. 0′. 7″. Now half this Angle is the Angle which these emergent Rays contain with the emergent mean refrangible Rays, and a quarter thereof, that is 30′. 2″. may be accounted the Angle which they would contain with the same emergent mean refrangible Rays, were they co-incident to them within the Glass, and suffered no other Refraction than that at their Emergence. For, if two equal Refractions, the one at the Incidence of the Rays on the Prism, the other at their Emergence, make half the Angle 2 deg. 0′. 7″. then one of those Refractions will make about a quarter of that Angle, and this quarter added to, and subducted from the Angle of Refraction of the mean refrangible Rays, which was 53 deg. 35′, gives the Angles of Refraction of the most and least refrangible Rays 54 deg. 5′ 2″, and 53 deg. 4′ 58″, whose Sines are 8099 and 7995, the common Angle of Incidence being 31 deg. 15′, and its Sine 5188; and these Sines in the least round Numbers are in proportion to one another, as 78 and 77 to 50.

Now, if you subduct the common Sine of Incidence 50 from the Sines of Refraction 77 and 78, the Remainders 27 and 28 shew, that in small Refractions the Refraction of the least refrangible Rays is to the Refraction of the most refrangible ones, as 27 to 28 very nearly, and that the difference of the Refractions of the least refrangible and most refrangible Rays is about the $27\frac{1}{2}$th Part of the whole Refraction of the mean refrangible Rays.

Whence they that are skilled in Opticks will easily understand,* that the Breadth of the least circular Space, into which Object-glasses of Telescopes can collect all sorts of Parallel Rays, is about the $27\frac{1}{2}$th Part of half the Aperture of the Glass, or 55th Part of the whole Aperture; and that the Focus of the most refrangible Rays is nearer to the Object-glass than the Focus of the least refrangible ones, by about the $27\frac{1}{2}$th Part of the distance between the Object-glass and the Focus of the mean refrangible ones.

And if Rays of all sorts, flowing from any one lucid Point in the Axis of any convex Lens, be made by the Refraction of the Lens to converge to Points not too remote from the Lens, the Focus of the most refrangible Rays shall be nearer to the Lens than the Focus of the least refrangible ones, by a distance which is to the $27\frac{1}{2}$th Part of the distance of the Focus of the mean refrangible Rays from the Lens, as the distance between that Focus and the lucid Point, from whence the Rays flow, is to the distance between that lucid Point and the Lens very nearly.

Now to examine whether the Difference between the Refractions, which the most refrangible and the least refrangible Rays flowing from the same Point suffer in the Object-glasses of Telescopes and such-like Glasses, be so great as is here described, I contrived the following Experiment.

* *This is demonstrated in our* Author's Lect. Optic. *Part* I. *Sect.* IV. *Prop.* 37.

Exper. 16. The Lens which I used in the second
and eighth Experiments, being placed six Feet and
an Inch distant from any Object, collected the
Species of that Object by the mean refrangible Rays
at the distance of six Feet and an Inch from the Lens
on the other side. And therefore by the foregoing
Rule, it ought to collect the Species of that Object
by the least refrangible Rays at the distance of six
Feet and $3\frac{2}{3}$ Inches from the Lens, and by the most
refrangible ones at the distance of five Feet and $10\frac{1}{3}$
Inches from it: So that between the two Places,
where these least and most refrangible Rays collect
the Species, there may be the distance of about $5\frac{1}{3}$
Inches. For by that Rule, as six Feet and an Inch
(the distance of the Lens from the lucid Object) is
to twelve Feet and two Inches (the distance of the
lucid Object from the Focus of the mean refrangible
Rays) that is, as One is to Two; so is the $27\frac{1}{2}$th Part
of six Feet and an Inch (the distance between the
Lens and the same Focus) to the distance between
the Focus of the most refrangible Rays and the Focus
of the least refrangible ones, which is therefore $5\frac{1}{5}\frac{7}{5}$
Inches, that is very nearly $5\frac{1}{3}$ Inches. Now to know
whether this Measure was true, I repeated the
second and eighth Experiment with coloured Light,
which was less compounded than that I there made
use of: For I now separated the heterogeneous Rays
from one another by the Method I described in the
eleventh Experiment, so as to make a coloured Spectrum about twelve or fifteen Times longer than
broad. This Spectrum I cast on a printed Book, and

placing the above-mentioned Lens at the distance of
six Feet and an Inch from this Spectrum to collect
the Species of the illuminated Letters at the same
distance on the other side, I found that the Species
of the Letters illuminated with blue were nearer to
the Lens than those illuminated with deep red by
about three Inches, or three and a quarter; but the
Species of the Letters illuminated with indigo and
violet appeared so confused and indistinct, that I
could not read them: Whereupon viewing the Prism,
I found it was full of Veins running from one end of
the Glass to the other; so that the Refraction could
not be regular. I took another Prism therefore
which was free from Veins, and instead of the
Letters I used two or three Parallel black Lines a
little broader than the Strokes of the Letters, and
casting the Colours upon these Lines in such manner,
that the Lines ran along the Colours from one end
of the Spectrum to the other, I found that the Focus
where the indigo, or confine of this Colour and violet
cast the Species of the black Lines most distinctly,
to be about four Inches, or $4\frac{1}{4}$ nearer to the Lens
than the Focus, where the deepest red cast the
Species of the same black Lines most distinctly.
The violet was so faint and dark, that I could not
discern the Species of the Lines distinctly by that
Colour; and therefore considering that the Prism was
made of a dark coloured Glass inclining to green, I
took another Prism of clear white Glass; but the
Spectrum of Colours which this Prism made had
long white Streams of faint Light shooting out from

both ends of the Colours, which made me conclude
that something was amiss; and viewing the Prism,
I found two or three little Bubbles in the Glass,
which refracted the Light irregularly. Wherefore I
covered that Part of the Glass with black Paper, and
letting the Light pass through another Part of it
which was free from such Bubbles, the Spectrum of
Colours became free from those irregular Streams of
Light, and was now such as I desired. But still I
found the violet so dark and faint, that I could
scarce see the Species of the Lines by the violet, and
not at all by the deepest Part of it, which was next
the end of the Spectrum. I suspected therefore, that
this faint and dark Colour might be allayed by that
scattering Light which was refracted, and reflected
irregularly, partly by some very small Bubbles in
the Glasses, and partly by the Inequalities of their
Polish; which Light, tho' it was but little, yet it being
of a white Colour, might suffice to affect the Sense
so strongly as to disturb the Phænomena of that
weak and dark Colour the violet, and therefore I
tried, as in the 12th, 13th, and 14th Experiments,
whether the Light of this Colour did not consist of
a sensible Mixture of heterogeneous Rays, but found
it did not. Nor did the Refractions cause any other
sensible Colour than violet to emerge out of this
Light, as they would have done out of white Light,
and by consequence out of this violet Light had it
been sensibly compounded with white Light. And
therefore I concluded, that the reason why I could
not see the Species of the Lines distinctly by this

Colour, was only the Darkness of this Colour, and Thinness of its Light, and its distance from the Axis of the Lens; I divided therefore those Parallel black Lines into equal Parts, by which I might readily know the distances of the Colours in the Spectrum from one another, and noted the distances of the Lens from the Foci of such Colours, as cast the Species of the Lines distinctly, and then considered whether the difference of those distances bear such proportion to $5\frac{1}{3}$ Inches, the greatest Difference of the distances, which the Foci of the deepest red and violet ought to have from the Lens, as the distance of the observed Colours from one another in the Spectrum bear to the greatest distance of the deepest red and violet measured in the Rectilinear Sides of the Spectrum, that is, to the Length of those Sides, or Excess of the Length of the Spectrum above its Breadth. And my Observations were as follows.

When I observed and compared the deepest sensible red, and the Colour in the Confine of green and blue, which at the Rectilinear Sides of the Spectrum was distant from it half the Length of those Sides, the Focus where the Confine of green and blue cast the Species of the Lines distinctly on the Paper, was nearer to the Lens than the Focus, where the red cast those Lines distinctly on it by about $2\frac{1}{2}$ or $2\frac{3}{4}$ Inches. For sometimes the Measures were a little greater, sometimes a little less, but seldom varied from one another above $\frac{1}{3}$ of an Inch. For it was very difficult to define the Places of the Foci, without some little Errors. Now, if the Colours distant half the Length of the Image, (measured at its Rectilinear Sides) give $2\frac{1}{2}$ or $2\frac{3}{4}$ Difference of the distances of their Foci from the Lens, then the Colours distant the whole Length ought to give 5 or $5\frac{1}{2}$ Inches difference of those distances.

But here it's to be noted, that I could not see the red to the full end of the Spectrum, but only to the Center of the Semicircle which bounded that end, or a little farther; and therefore I compared this red not with that Colour which was exactly in the middle of the Spectrum, or Confine of green and blue, but with that which verged a little more to the blue than to the green: And as I reckoned the whole Length of the Colours not to be the whole Length of the Spectrum, but the Length of its Rectilinear Sides, so compleating the semicircular Ends into Circles, when either of the observed Colours fell within those Circles, I measured the distance of that Colour from the semicircular End of the Spectrum, and subducting half this distance from the measured distance of the two Colours, I took the Remainder for their corrected distance; and in these Observations set down this corrected distance for the difference of the distances of their Foci from the Lens. For, as the Length of the Rectilinear Sides of the Spectrum would be the whole Length of all the Colours, were the Circles of which (as we shewed) that Spectrum consists contracted and reduced to Physical Points, so in that Case this corrected distance would be the real distance of the two observed Colours.

When therefore I farther observed the deepest

sensible red, and that blue whose corrected distance from it was $\frac{7}{12}$ Parts of the Length of the Rectilinear Sides of the Spectrum, the difference of the distances of their Foci from the Lens was about $3\frac{1}{4}$ Inches, and as 7 to 12, so is $3\frac{1}{4}$ to $5\frac{4}{7}$.

When I observed the deepest sensible red, and that indigo whose corrected distance was $\frac{8}{12}$ or $\frac{2}{3}$ of the Length of the Rectilinear Sides of the Spectrum, the difference of the distances of their Foci from the Lens, was about $3\frac{2}{3}$ Inches, and as 2 to 3, so is $3\frac{2}{3}$ to $5\frac{1}{2}$.

When I observed the deepest sensible red, and that deep indigo whose corrected distance from one another was $\frac{9}{12}$ or $\frac{3}{4}$ of the Length of the Rectilinear Sides of the Spectrum, the difference of the distances of their Foci from the Lens was about 4 Inches; and as 3 to 4, so is 4 to $5\frac{1}{3}$.

When I observed the deepest sensible red, and that Part of the violet next the indigo, whose corrected distance from the red was $\frac{10}{12}$ or $\frac{5}{6}$ of the Length of the Rectilinear Sides of the Spectrum, the difference of the distances of their Foci from the Lens was about $4\frac{1}{2}$ Inches, and as 5 to 6, so is $4\frac{1}{2}$ to $5\frac{2}{5}$. For sometimes, when the Lens was advantageously placed, so that its Axis respected the blue, and all Things else were well ordered, and the Sun shone clear, and I held my Eye very near to the Paper on which the Lens cast the Species of the Lines, I could see pretty distinctly the Species of those Lines by that Part of the violet which was next the indigo; and sometimes I could see them by above half the violet,

For in making these Experiments I had observed, that the Species of those Colours only appear distinct, which were in or near the Axis of the Lens: So that if the blue or indigo were in the Axis, I could see their Species distinctly; and then the red appeared much less distinct than before. Wherefore I contrived to make the Spectrum of Colours shorter than before, so that both its Ends might be nearer to the Axis of the Lens. And now its Length was about $2\frac{1}{2}$ Inches, and Breadth about $\frac{1}{5}$ or $\frac{1}{6}$ of an Inch. Also instead of the black Lines on which the Spectrum was cast, I made one black Line broader than those, that I might see its Species more easily; and this Line I divided by short cross Lines into equal Parts, for measuring the distances of the observed Colours. And now I could sometimes see the Species of this Line with its Divisions almost as far as the Center of the semicircular violet End of the Spectrum, and made these farther Observations.

When I observed the deepest sensible red, and that Part of the violet, whose corrected distance from it was about $\frac{8}{9}$ Parts of the Rectilinear Sides of the Spectrum, the Difference of the distances of the Foci of those Colours from the Lens, was one time $4\frac{2}{3}$, another time $4\frac{3}{4}$, another time $4\frac{7}{8}$ Inches; and as 8 to 9, so are $4\frac{2}{3}$, $4\frac{3}{4}$, $4\frac{7}{8}$, to $5\frac{1}{4}$, $5\frac{11}{32}$, $5\frac{31}{64}$ respectively.

When I observed the deepest sensible red, and deepest sensible violet, (the corrected distance of which Colours, when all Things were ordered to the best Advantage, and the Sun shone very clear, was about $\frac{11}{12}$ or $\frac{15}{16}$ Parts of the Length of the Rectilinear

Sides of the coloured Spectrum) I found the Difference of the distances of their Foci from the Lens
sometimes $4\frac{3}{4}$ sometimes $5\frac{1}{4}$, and for the most part
5 Inches or thereabouts; and as 11 to 12, or 15 to 16,
so is five Inches to $5\frac{2}{3}$ or $5\frac{1}{3}$ Inches.

And by this Progression of Experiments I satisfied
my self, that had the Light at the very Ends of the
Spectrum been strong enough to make the Species
of the black Lines appear plainly on the Paper, the
Focus of the deepest violet would have been found
nearer to the Lens, than the Focus of the deepest red,
by about $5\frac{1}{3}$ Inches at least. And this is a farther
Evidence, that the Sines of Incidence and Refraction
of the several sorts of Rays, hold the same Proportion
to one another in the smallest Refractions which they
do in the greatest.

My Progress in making this nice and troublesome
Experiment I have set down more at large, that they
that shall try it after me may be aware of the Circumspection requisite to make it succeed well. And if
they cannot make it succeed so well as I did, they
may notwithstanding collect by the Proportion of
the distance of the Colours of the Spectrum, to the
Difference of the distances of their Foci from the
Lens, what would be the Success in the more distant
Colours by a better trial. And yet, if they use a
broader Lens than I did, and fix it to a long strait
Staff, by means of which it may be readily and truly
directed to the Colour whose Focus is desired, I
question not but the Experiment will succeed better
with them than it did with me. For I directed the

Axis as nearly as I could to the middle of the Colours, and then the faint Ends of the Spectrum being remote from the Axis, cast their Species less distinctly on the Paper than they would have done, had the Axis been successively directed to them.

Now by what has been said, it's certain that the Rays which differ in Refrangibility do not converge to the same Focus; but if they flow from a lucid Point, as far from the Lens on one side as their Foci are on the other, the Focus of the most refrangible Rays shall be nearer to the Lens than that of the least refrangible, by above the fourteenth Part of the whole distance; and if they flow from a lucid Point, so very remote from the Lens, that before their Incidence they may be accounted parallel, the Focus of the most refrangible Rays shall be nearer to the Lens than the Focus of the least refrangible, by about the 27th or 28th Part of their whole distance from it. And the Diameter of the Circle in the middle Space between those two Foci which they illuminate, when they fall there on any Plane, perpendicular to the Axis (which Circle is the least into which they can all be gathered) is about the 55th Part of the Diameter of the Aperture of the Glass. So that 'tis a wonder, that Telescopes represent Objects so distinct as they do. But were all the Rays of Light equally refrangible, the Error arising only from the Sphericalness of the Figures of Glasses would be many hundred times less. For, if the Object-glass of a Telescope be Plano-convex, and the Plane side be turned towards the Object, and the Diameter of the

Sphere, whereof this Glass is a Segment, be called D, and the Semidiameter of the Aperture of the Glass be called S, and the Sine of Incidence out of Glass into Air, be to the Sine of Refraction as I to R; the Rays which come parallel to the Axis of the Glass, shall in the Place where the Image of the Object is most distinctly made, be scattered all over a little Circle, whose Diameter is $\dfrac{Rq}{Iq} \times \dfrac{S\ cub.}{D\ quad.}$ very nearly,* as I gather by computing the Errors of the Rays by the Method of infinite Series, and rejecting the Terms, whose Quantities are inconsiderable. As for instance, if the Sine of Incidence I, be to the Sine of Refraction R, as 20 to 31, and if D the Diameter of the Sphere, to which the Convex-side of the Glass is ground, be 100 Feet or 1200 Inches, and S the Semidiameter of the Aperture be two Inches, the Diameter of the little Circle, $\left(\text{that is } \dfrac{Rq \times S\ cub.}{Iq \times D\ quad.}\right)$ will be $\dfrac{31 \times 31 \times 8}{20 \times 20 \times 1200 \times 1200}$ (or $\dfrac{961}{72000000}$) Parts of an Inch. But the Diameter of the little Circle, through which these Rays are scattered by unequal Refrangibility, will be about the 55th Part of the Aperture of the Object-glass, which here is four Inches. And therefore, the Error arising from the Spherical Figure of the Glass, is to the Error arising from the different Refrangibility of the Rays, as $\dfrac{961}{72000000}$ to $\dfrac{4}{55}$, that is as 1 to 5449; and therefore

* *How to do this, is shewn in our* Author's Lect. Optic. *Part* I. *Sect.* IV. *Prop.* 31.

being in comparison so very little, deserves not to be considered.

But you will say, if the Errors caused by the different Refrangibility be so very great, how comes it to pass, that Objects appear through Telescopes so distinct as they do? I answer, 'tis because the erring Rays are not scattered uniformly over all that Circular Space, but collected infinitely more densely in the Center than in any other Part of the Circle,

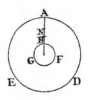

Fig. 27.

and in the Way from the Center to the Circumference, grow continually rarer and rarer, so as at the Circumference to become infinitely rare; and by reason of their Rarity are not strong enough to be visible, unless in the Center and very near it. Let ADE [in *Fig*. 27.] represent one of those Circles described with the Center C, and Semidiameter AC, and let BFG be a smaller Circle concentrick to the former, cutting with its Circumference the Diameter AC in B, and bisect AC in N; and by my reckoning, the Density of the Light in any Place B, will be to its Density in N, as AB to BC; and the whole Light within the lesser Circle BFG, will be to the whole Light within the greater AED, as the Excess of the

Square of AC above the Square of AB, is to the Square of AC. As if BC be the fifth Part of AC, the Light will be four times denser in B than in N, and the whole Light within the less Circle, will be to the whole Light within the greater, as nine to twenty-five. Whence it's evident, that the Light within the less Circle, must strike the Sense much more strongly, than that faint and dilated Light round about between it and the Circumference of the greater.

But it's farther to be noted, that the most luminous of the Prismatick Colours are the yellow and orange. These affect the Senses more strongly than all the rest together, and next to these in strength are the red and green. The blue compared with these is a faint and dark Colour, and the indigo and violet are much darker and fainter, so that these compared with the stronger Colours are little to be regarded. The Images of Objects are therefore to be placed, not in the Focus of the mean refrangible Rays, which are in the Confine of green and blue, but in the Focus of those Rays which are in the middle of the orange and yellow; there where the Colour is most luminous and fulgent, that is in the brightest yellow, that yellow which inclines more to orange than to green. And by the Refraction of these Rays (whose Sines of Incidence and Refraction in Glass are as 17 and 11) the Refraction of Glass and Crystal for Optical Uses is to be measured. Let us therefore place the Image of the Object in the Focus of these Rays, and all the yellow and orange will fall within a

Circle, whose Diameter is about the 250th Part of
the Diameter of the Aperture of the Glass. And if
you add the brighter half of the red, (that half which
is next the orange) and the brighter half of the green,
(that half which is next the yellow) about three fifth
Parts of the Light of these two Colours will fall within
the same Circle, and two fifth Parts will fall without it round about; and that which falls without will
be spread through almost as much more space as
that which falls within, and so in the gross be almost
three times rarer. Of the other half of the red and
green, (that is of the deep dark red and willow green)
about one quarter will fall within this Circle, and
three quarters without, and that which falls without
will be spread through about four or five times more
space than that which falls within; and so in the
gross be rarer, and if compared with the whole Light
within it, will be about 25 times rarer than all that
taken in the gross; or rather more than 30 or 40 times
rarer, because the deep red in the end of the Spectrum
of Colours made by a Prism is very thin and rare,
and the willow green is something rarer than the
orange and yellow. The Light of these Colours therefore being so very much rarer than that within the
Circle, will scarce affect the Sense, especially since the
deep red and willow green of this Light, are much
darker Colours than the rest. And for the same reason
the blue and violet being much darker Colours than
these, and much more rarified, may be neglected.
For the dense and bright Light of the Circle, will
obscure the rare and weak Light of these dark

Colours round about it, and render them almost insensible. The sensible Image of a lucid Point is therefore scarce broader than a Circle, whose Diameter is
the 250th Part of the Diameter of the Aperture of
the Object-glass of a good Telescope, or not much
broader, if you except a faint and dark misty Light
round about it, which a Spectator will scarce regard.
And therefore in a Telescope, whose Aperture is
four Inches, and Length an hundred Feet, it exceeds
not 2″ 45‴, or 3″. And in a Telescope whose Aperture is two Inches, and Length 20 or 30 Feet, it may
be 5″ or 6″, and scarce above. And this answers well
to Experience: For some Astronomers have found
the Diameters of the fix'd Stars, in Telescopes of
between 20 and 60 Feet in length, to be about 5″ or
6″, or at most 8″ or 10″ in diameter. But if the Eye-
Glass be tincted faintly with the Smoak of a Lamp
or Torch, to obscure the Light of the Star, the fainter
Light in the Circumference of the Star ceases to be
visible, and the Star (if the Glass be sufficiently
soiled with Smoak) appears something more like a
mathematical Point. And for the same Reason, the
enormous Part of the Light in the Circumference of
every lucid Point ought to be less discernible in
shorter Telescopes than in longer, because the
shorter transmit less Light to the Eye.

Now, that the fix'd Stars, by reason of their im-
mense Distance, appear like Points, unless so far as
their Light is dilated by Refraction, may appear from
hence; that when the Moon passes over them and
eclipses them, their Light vanishes, not gradually

like that of the Planets, but all at once; and in the end of the Eclipse it returns into Sight all at once, or certainly in less time than the second of a Minute; the Refraction of the Moon's Atmosphere a little protracting the time in which the Light of the Star first vanishes, and afterwards returns into Sight.

Now, if we suppose the sensible Image of a lucid Point, to be even 250 times narrower than the Aperture of the Glass; yet this Image would be still much greater than if it were only from the spherical Figure of the Glass. For were it not for the different Refrangibility of the Rays, its breadth in an 100 Foot Telescope whose aperture is 4 Inches, would be but $\frac{961}{72000000}$ parts of an Inch, as is manifest by the foregoing Computation. And therefore in this case the greatest Errors arising from the spherical Figure of the Glass, would be to the greatest sensible Errors arising from the different Refrangibility of the Rays as $\frac{961}{72000000}$ to $\frac{4}{250}$ at most, that is only as 1 to 1200. And this sufficiently shews that it is not the spherical Figures of Glasses, but the different Refrangibility of the Rays which hinders the perfection of Telescopes.

There is another Argument by which it may appear that the different Refrangibility of Rays, is the true cause of the imperfection of Telescopes. For the Errors of the Rays arising from the spherical Figures of Object-glasses, are as the Cubes of the Apertures of the Object Glasses; and thence to make Telescopes of various Lengths magnify with equal distinctness, the Apertures of the Object-glasses, and the Charges or magnifying Powers ought to be as the

Cubes of the square Roots of their lengths; which doth not answer to Experience. But the Errors of the Rays arising from the different Refrangibility, are as the Apertures of the Object-glasses; and thence to make Telescopes of various lengths, magnify with equal distinctness, their Apertures and Charges ought to be as the square Roots of their lengths; and

Fig. 28.

this answers to Experience, as is well known. For Instance, a Telescope of 64 Feet in length, with an Aperture of $2\frac{2}{3}$ Inches, magnifies about 120 times, with as much distinctness as one of a Foot in length, with $\frac{1}{3}$ of an Inch aperture, magnifies 15 times.

Now were it not for this different Refrangibility of Rays, Telescopes might be brought to a greater perfection than we have yet describ'd, by composing the Object-glass of two Glasses with Water between them. Let ADFC [in *Fig*. 28.] represent the Object-glass composed of two Glasses ABED and BEFC, alike convex on the outsides AGD and CHF, and alike concave on the insides BME, BNE, with Water in the concavity BMEN. Let the Sine of Incidence

out of Glass into Air be as I to R, and out of Water
into Air, as K to R, and by consequence out of Glass
into Water, as I to K: and let the Diameter of the
Sphere to which the convex sides AGD and CHF are
ground be D, and the Diameter of the Sphere to
which the concave sides BME and BNE, are ground
be to D, as the Cube Root of KK—KI to the Cube
Root of RK—RI: and the Refractions on the concave
sides of the Glasses, will very much correct the
Errors of the Refractions on the convex sides, so far
as they arise from the sphericalness of the Figure.
And by this means might Telescopes be brought to
sufficient perfection, were it not for the different Refrangibility of several sorts of Rays. But by reason of
this different Refrangibility, I do not yet see any
other means of improving Telescopes by Refractions
alone, than that of increasing their lengths, for which
end the late Contrivance of *Hugenius* seems well
accommodated. For very long Tubes are cumber-
some, and scarce to be readily managed, and by
reason of their length are very apt to bend, and shake
by bending, so as to cause a continual trembling in
the Objects, whereby it becomes difficult to see them
distinctly: whereas by his Contrivance the Glasses
are readily manageable, and the Object-glass being
fix'd upon a strong upright Pole becomes more
steady.

    Seeing therefore the Improvement of Telescopes
of given lengths by Refractions is desperate; I contrived heretofore a Perspective by Reflexion, using
instead of an Object-glass a concave Metal. The dia-

meter of the Sphere to which the Metal was ground
concave was about 25 *English* Inches, and by consequence the length of the Instrument about six Inches
and a quarter. The Eye-glass was Plano-convex, and
the diameter of the Sphere to which the convex side
was ground was about $\frac{1}{5}$ of an Inch, or a little less,
and by consequence it magnified between 30 and 40
times. By another way of measuring I found that it
magnified about 35 times. The concave Metal bore
an Aperture of an Inch and a third part; but the
Aperture was limited not by an opake Circle, covering the Limb of the Metal round about, but by an
opake Circle placed between the Eyeglass and the
Eye, and perforated in the middle with a little round
hole for the Rays to pass through to the Eye. For this
Circle by being placed here, stopp'd much of the
erroneous Light, which otherwise would have disturbed the Vision. By comparing it with a pretty
good Perspective of four Feet in length, made with
a concave Eye-glass, I could read at a greater distance
with my own Instrument than with the Glass. Yet
Objects appeared much darker in it than in the Glass,
and that partly because more Light was lost by Reflexion in the Metal, than by Refraction in the Glass,
and partly because my Instrument was overcharged.
Had it magnified but 30 or 25 times, it would have
made the Object appear more brisk and pleasant.
Two of these I made about 16 Years ago, and have
one of them still by me, by which I can prove the
truth of what I write. Yet it is not so good as at the
first. For the concave has been divers times tarnished

and cleared again, by rubbing it with very soft Leather. When I made these an Artist in *London* undertook to imitate it; but using another way of polishing them than I did, he fell much short of what I had attained to, as I afterwards understood by discoursing the Under-workman he had employed. The Polish I used was in this manner. I had two round Copper Plates, each six Inches in Diameter, the one convex, the other concave, ground very true to one another. On the convex I ground the Object-Metal or Concave which was to be polish'd, 'till it had taken the Figure of the Convex and was ready for a Polish. Then I pitched over the convex very thinly, by dropping melted Pitch upon it, and warming it to keep the Pitch soft, whilst I ground it with the concave Copper wetted to make it spread eavenly all over the convex. Thus by working it well I made it as thin as a Groat, and after the convex was cold I ground it again to give it as true a Figure as I could. Then I took Putty which I had made very fine by washing it from all its grosser Particles, and laying a little of this upon the Pitch, I ground it upon the Pitch with the concave Copper, till it had done making a Noise; and then upon the Pitch I ground the Object-Metal with a brisk motion, for about two or three Minutes of time, leaning hard upon it. Then I put fresh Putty upon the Pitch, and ground it again till it had done making a noise, and afterwards ground the Object-Metal upon it as before. And this Work I repeated till the Metal was polished, grinding it the last time with all my strength for a good while together, and frequently breathing upon the Pitch, to keep it moist without laying on any more fresh Putty. The Object-Metal was two Inches broad, and about one third part of an Inch thick, to keep it from bending. I had two of these Metals, and when I had polished them both, I tried which was best, and ground the other again, to see if I could make it better than that which I kept. And thus by many Trials I learn'd the way of polishing, till I made those two reflecting Perspectives I spake of above. For this Art of polishing will be better learn'd by repeated Practice than by my Description. Before I ground the Object-Metal on the Pitch, I always ground the Putty on it with the concave Copper, till it had done making a noise, because if the Particles of the Putty were not by this means made to stick fast in the Pitch, they would by rolling up and down grate and fret the Object-Metal and fill it full of little holes.

But because Metal is more difficult to polish than Glass, and is afterwards very apt to be spoiled by tarnishing, and reflects not so much Light as Glass quick-silver'd over does: I would propound to use instead of the Metal, a Glass ground concave on the foreside, and as much convex on the back-side, and quick-silver'd over on the convex side. The Glass must be every where of the same thickness exactly. Otherwise it will make Objects look colour'd and indistinct. By such a Glass I tried about five or six Years ago to make a reflecting Telescope of four Feet in length to magnify about 150 times, and I satisfied my self that there wants nothing but a good

Artist to bring the Design to perfection. For the Glass being wrought by one of our *London* Artists after such a manner as they grind Glasses for Telescopes, though it seemed as well wrought as the Object-glasses use to be, yet when it was quicksilver'd, the Reflexion discovered innumerable Inequalities all over the Glass. And by reason of these Inequalities, Objects appeared indistinct in this Instrument. For the Errors of reflected Rays caused by any Inequality of the Glass, are about six times greater than the Errors of refracted Rays caused by the like Inequalities. Yet by this Experiment I satisfied my self that the Reflexion on the concave side of the Glass, which I feared would disturb the Vision, did no sensible prejudice to it, and by consequence that nothing is wanting to perfect these Telescopes, but good Workmen who can grind and polish Glasses truly spherical. An Object-glass of a fourteen Foot Telescope, made by an Artificer at *London*, I once mended considerably, by grinding it on Pitch with Putty, and leaning very easily on it in the grinding, lest the Putty should scratch it. Whether this way may not do well enough for polishing these reflecting Glasses, I have not yet tried. But he that shall try either this or any other way of polishing which he may think better, may do well to make his Glasses ready for polishing, by grinding them without that Violence, wherewith our *London* Workmen press their Glasses in grinding. For by such violent pressure, Glasses are apt to bend a little in the grinding, and such bending will certainly spoil their

Figure. To recommend therefore the consideration of these reflecting Glasses to such Artists as are curious in figuring Glasses, I shall describe this optical Instrument in the following Proposition.

PROP. VIII. PROB. II.

To shorten Telescopes.

LET ABCD [in *Fig.* 29.] represent a Glass spherically concave on the foreside AB, and as much convex on the backside CD, so that it be every where of an equal thickness. Let it not be thicker on one side than on the other, lest it make Objects appear colour'd and indistinct, and let it be very truly wrought and quick-silver'd over on the backside; and set in the Tube VXYZ which must be very black within. Let EFG represent a Prism of Glass or Crystal placed near the other end of the Tube, in the middle of it, by means of a handle of Brass or Iron FGK, to the end of which made flat it is cemented. Let this Prism be rectangular at E, and let the other two Angles at F and G be accurately equal to each other, and by consequence equal to half right ones, and let the plane sides FE and GE be square, and by consequence the third side FG a rectangular Parallelogram, whose length is to its breadth in a subduplicate proportion of two to one. Let it be so placed in the Tube, that the Axis of the Speculum may pass through the middle of the square side EF perpen-

dicularly and by consequence through the middle of
the side FG at an Angle of 45 Degrees, and let the
side EF be turned towards the Speculum, and the
distance of this Prism from the Speculum be such
that the Rays of the Light PQ, RS, &c. which are
incident upon the Speculum in Lines parallel to the
Axis thereof, may enter the Prism at the side EF, and
be reflected by the side FG, and thence go out of it
through the side GE, to the Point T, which must be
the common Focus of the Speculum ABDC, and of a
Plano-convex Eye-glass H, through which those
Rays must pass to the Eye. And let the Rays at their
coming out of the Glass pass through a small round
hole, or aperture made in a little plate of Lead, Brass,
or Silver, wherewith the Glass is to be covered,
which hole must be no bigger than is necessary for
Light enough to pass through. For so it will render
the Object distinct, the Plate in which 'tis made
intercepting all the erroneous part of the Light which
comes from the verges of the Speculum AB. Such an
Instrument well made, if it be six Foot long, (reckoning the length from the Speculum to the Prism, and
thence to the Focus T) will bear an aperture of six
Inches at the Speculum, and magnify between
two and three hundred times. But the hole H here
limits the aperture with more advantage, than if the
aperture was placed at the Speculum. If the Instrument be made longer or shorter, the aperture must
be in proportion as the Cube of the square-square
Root of the length, and the magnifying as the aperture. But it's convenient that the Speculum be an

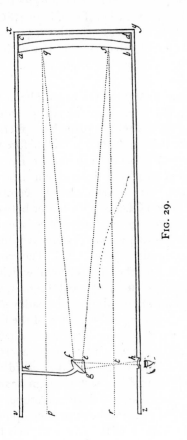

FIG. 29.

Inch or two broader than the aperture at the least, and that the Glass of the Speculum be thick, that it bend not in the working. The Prism EFG must be no bigger than is necessary, and its back side FG must not be quick-silver'd over. For without quicksilver it will reflect all the Light incident on it from the Speculum.

In this Instrument the Object will be inverted, but may be erected by making the square sides FF and EG of the Prism EFG not plane but spherically convex, that the Rays may cross as well before they come at it as afterwards between it and the Eye-glass. If it be desired that the Instrument bear a larger aperture, that may be also done by composing the Speculum of two Glasses with Water between them.

If the Theory of making Telescopes could at length be fully brought into Practice, yet there would be certain Bounds beyond which Telescopes could not perform. For the Air through which we look upon the Stars, is in a perpetual Tremor; as may be seen by the tremulous Motion of Shadows cast from high Towers, and by the twinkling of the fix'd Stars. But these Stars do not twinkle when viewed through Telescopes which have large apertures. For the Rays of Light which pass through divers parts of the aperture, tremble each of them apart, and by means of their various and sometimes contrary Tremors, fall at one and the same time upon different points in the bottom of the Eye, and their trembling Motions are too quick and confused to be perceived severally. And all these illuminated Points constitute one broad lucid Point, composed of those many trembling Points confusedly and insensibly mixed with one another by very short and swift Tremors, and thereby cause the Star to appear broader than it is, and without any trembling of the whole. Long Telescopes may cause Objects to appear brighter and larger than short ones can do, but they cannot be so formed as to take away that confusion of the Rays which arises from the Tremors of the Atmosphere. The only Remedy is a most serene and quiet Air, such as may perhaps be found on the tops of the highest Mountains above the grosser Clouds.

THE

FIRST BOOK

OF

OPTICKS

PART II.

PROP. I. THEOR. I.

The Phænomena of Colours in refracted or reflected Light are not caused by new Modifications of the Light variously impress'd, according to the various Terminations of the Light and Shadow.

The PROOF by Experiments.

Exper 1. FOR if the Sun shine into a very dark Chamber through an oblong hole F, [in *Fig.* 1.] whose breadth is the sixth or eighth part of an Inch, or something less; and his beam FH do afterwards pass first through a very large Prism ABC, distant about 20 Feet from the hole, and parallel to it, and then (with its white part) through an oblong hole H, whose breadth is about the fortieth or sixtieth

part of an Inch, and which is made in a black opake
Body GI, and placed at the distance of two or three
Feet from the Prism, in a parallel Situation both to
the Prism and to the former hole, and if this white
Light thus transmitted through the hole H, fall
afterwards upon a white Paper pt, placed after that
hole H, at the distance of three or four Feet from it,
and there paint the usual Colours of the Prism,
suppose red at t, yellow at s, green at r, blue at q,
and violet at p; you may with an Iron Wire, or any
such like slender opake Body, whose breadth is about
the tenth part of an Inch, by intercepting the Rays
at k, l, m, n or o, take away any one of the Colours at
t, s, r, q or p, whilst the other Colours remain upon
the Paper as before; or with an Obstacle something
bigger you may take away any two, or three, or four
Colours together, the rest remaining: So that any
one of the Colours as well as violet may become outmost in the Confine of the Shadow towards p, and
any one of them as well as red may become outmost
in the Confine of the Shadow towards t, and any
one of them may also border upon the Shadow made
within the Colours by the Obstacle R intercepting
some intermediate part of the Light; and, lastly, any
one of them by being left alone, may border upon
the Shadow on either hand. All the Colours have
themselves indifferently to any Confines of Shadow,
and therefore the differences of these Colours from
one another, do not arise from the different Confines
of Shadow, whereby Light is variously modified, as
has hitherto been the Opinion of Philosophers. In

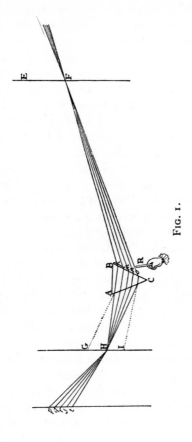

FIG. 1.

trying these things 'tis to be observed, that by how much the holes F and H are narrower, and the Intervals between them and the Prism greater, and the Chamber darker, by so much the better doth the Experiment succeed; provided the Light be not so far diminished, but that the Colours at *pt* be sufficiently visible. To procure a Prism of solid Glass large enough for this Experiment will be difficult, and therefore a prismatick Vessel must be made of polish'd Glass Plates cemented together, and filled with salt Water or clear Oil.

Exper. 2. The Sun's Light let into a dark Chamber through the round hole F, [in *Fig.* 2.] half an Inch wide, passed first through the Prism ABC placed at the hole, and then through a Lens PT something more than four Inches broad, and about eight Feet distant from the Prism, and thence converged to O the Focus of the Lens distant from it about three Feet, and there fell upon a white Paper DE. If that Paper was perpendicular to that Light incident upon it, as 'tis represented in the posture DE, all the Colours upon it at O appeared white. But if the Paper being turned about an Axis parallel to the Prism, became very much inclined to the Light, as 'tis represented in the Positions *de* and $\delta\varepsilon$; the same Light in the one case appeared yellow and red, in the other blue. Here one and the same part of the Light in one and the same place, according to the various Inclinations of the Paper, appeared in one case white, in another yellow or red, in a third blue, whilst the Confine of Light and shadow, and the

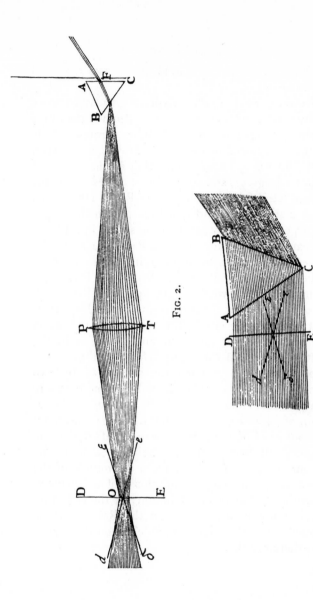

FIG. 2.

FIG. 3.

Refractions of the Prism in all these cases remained the same.

Exper. 3. Such another Experiment may be more easily tried as follows. Let a broad beam of the Sun's Light coming into a dark Chamber through a hole in the Window-shut be refracted by a large Prism ABC, [in *Fig.* 3.] whose refracting Angle C is more than 60 Degrees, and so soon as it comes out of the Prism, let it fall upon the white Paper DE glewed upon a stiff Plane; and this Light, when the Paper is perpendicular to it, as 'tis represented in DE, will appear perfectly white upon the Paper; but when the Paper is very much inclin'd to it in such a manner as to keep always parallel to the Axis of the Prism, the whiteness of the whole Light upon the Paper will according to the inclination of the Paper this way or that way, change either into yellow and red, as in the posture *de*, or into blue and violet, as in the posture $\delta\varepsilon$. And if the Light before it fall upon the Paper be twice refracted the same way by two parallel Prisms, these Colours will become the more conspicuous. Here all the middle parts of the broad beam of white Light which fell upon the Paper, did without any Confine of Shadow to modify it, become colour'd all over with one uniform Colour, the Colour being always the same in the middle of the Paper as at the edges, and this Colour changed according to the various Obliquity of the reflecting Paper, without any change in the Refractions or Shadow, or in the Light which fell upon the Paper. And therefore these Colours are to be derived from some other Cause than the new Modifications of Light by Refractions and Shadows.

If it be asked, what then is their Cause? I answer, That the Paper in the posture *de*, being more oblique to the more refrangible Rays than to the less refrangible ones, is more strongly illuminated by the latter than by the former, and therefore the less refrangible Rays are predominant in the reflected Light. And where-ever they are predominant in any Light, they tinge it with red or yellow, as may in some measure appear by the first Proposition of the first Part of this Book, and will more fully appear hereafter. And the contrary happens in the posture of the Paper δε, the more refrangible Rays being then predominant which always tinge Light with blues and violets.

Exper. 4. The Colours of Bubbles with which Children play are various, and change their Situation variously, without any respect to any Confine or Shadow. If such a Bubble be cover'd with a concave Glass, to keep it from being agitated by any Wind or Motion of the Air, the Colours will slowly and regularly change their situation, even whilst the Eye and the Bubble, and all Bodies which emit any Light, or cast any Shadow, remain unmoved. And therefore their Colours arise from some regular Cause which depends not on any Confine of Shadow. What this Cause is will be shewed in the next Book.

To these Experiments may be added the tenth Experiment of the first Part of this first Book, where the Sun's Light in a dark Room being trajected through the parallel Superficies of two Prisms tied together in the form of a Parallelopipede, became totally of one uniform yellow or red Colour, at its emerging out of the Prisms. Here, in the production of these Colours, the Confine of Shadow can have nothing to do. For the Light changes from white to yellow, orange and red successively, without any alteration of the Confine of Shadow: And at both edges of the emerging Light where the contrary Confines of Shadow ought to produce different Effects, the Colour is one and the same, whether it be white, yellow, orange or red: And in the middle of the emerging Light, where there is no Confine of Shadow at all, the Colour is the very same as at the edges, the whole Light at its very first Emergence being of one uniform Colour, whether white, yellow, orange or red, and going on thence perpetually without any change of Colour, such as the Confine of Shadow is vulgarly supposed to work in refracted Light after its Emergence. Neither can these Colours arise from any new Modifications of the Light by Refractions, because they change successively from white to yellow, orange and red, while the Refractions remain the same, and also because the Refractions are made contrary ways by parallel Superficies which destroy one another's Effects. They arise not therefore from any Modifications of Light made by Refractions and Shadows, but have some other Cause. What that Cause is we shewed above in this tenth Experiment, and need not here repeat it.

There is yet another material Circumstance of this Experiment. For this emerging Light being by a third Prism HIK [in *Fig*. 22. *Part* I.]* refracted towards the Paper PT, and there painting the usual Colours of the Prism, red, yellow, green, blue, violet: If these Colours arose from the Refractions of that Prism modifying the Light, they would not be in the Light before its Incidence on that Prism. And yet in that Experiment we found, that when by turning the two first Prisms about their common Axis all the Colours were made to vanish but the red; the Light which makes that red being left alone, appeared of the very same red Colour before its Incidence on the third Prism. And in general we find by other Experiments, that when the Rays which differ in Refrangibility are separated from one another, and any one Sort of them is considered apart, the Colour of the Light which they compose cannot be changed by any Refraction or Reflexion whatever, as it ought to be were Colours nothing else than Modifications of Light caused by Refractions, and Reflexions, and Shadows. This Unchangeableness of Colour I am now to describe in the following Proposition.

*See p. 59.

PROP. II.   Theor. II.

*All homogeneal Light has its proper Colour answering
to its Degree of Refrangibility, and that Colour
cannot be changed by Reflexions and Refractions.*

IN the Experiments of the fourth Proposition of
the first Part of this first Book, when I had separated the heterogeneous Rays from one another, the
Spectrum *pt* formed by the separated Rays, did in
the Progress from its End *p*, on which the most refrangible Rays fell, unto its other End *t*, on which the
least refrangible Rays fell, appear tinged with this
Series of Colours, violet, indigo, blue, green, yellow,
orange, red, together with all their intermediate Degrees in a continual Succession perpetually varying.
So that there appeared as many Degrees of Colours,
as there were sorts of Rays differing in Refrangibility.

Exper. 5. Now, that these Colours could not be
changed by Refraction, I knew by refracting with a
Prism sometimes one very little Part of this Light,
sometimes another very little Part, as is described in
the twelfth Experiment of the first Part of this Book.
For by this Refraction the Colour of the Light was
never changed in the least. If any Part of the red
Light was refracted, it remained totally of the same
red Colour as before. No orange, no yellow, no green
or blue, no other new Colour was produced by that
Refraction. Neither did the Colour any ways change
by repeated Refractions, but continued always the

same red entirely as at first. The like Constancy and
Immutability I found also in the blue, green, and
other Colours. So also, if I looked through a Prism
upon any Body illuminated with any part of this
homogeneal Light, as in the fourteenth Experiment
of the first Part of this Book is described; I could not
perceive any new Colour generated this way. All
Bodies illuminated with compound Light appear
through Prisms confused, (as was said above) and
tinged with various new Colours, but those illuminated with homogeneal Light appeared through
Prisms neither less distinct, nor otherwise colour'd,
than when viewed with the naked Eyes. Their
Colours were not in the least changed by the Refraction of the interposed Prism. I speak here of a
sensible Change of Colour: For the Light which I
here call homogeneal, being not absolutely homogeneal, there ought to arise some little Change of
Colour from its Heterogeneity. But, if that Heterogeneity was so little as it might be made by the said
Experiments of the fourth Proposition, that Change
was not sensible, and therefore in Experiments,
where Sense is Judge, ought to be accounted none
at all.

Exper. 6. And as these Colours were not change-
able by Refractions, so neither were they by Reflexions. For all white, grey, red, yellow, green,
blue, violet Bodies, as Paper, Ashes, red Lead, Orpiment, Indico Bise, Gold, Silver, Copper, Grass, blue
Flowers, Violets, Bubbles of Water tinged with
various Colours, Peacock's Feathers, the Tincture of

Lignum Nephriticum, and such-like, in red homogeneal Light appeared totally red, in blue Light
totally blue, in green Light totally green, and so of
other Colours. In the homogeneal Light of any
Colour they all appeared totally of that same Colour,
with this only Difference, that some of them reflected that Light more strongly, others more faintly.
I never yet found any Body, which by reflecting
homogeneal Light could sensibly change its Colour.

From all which it is manifest, that if the Sun's
Light consisted of but one sort of Rays, there would
be but one Colour in the whole World, nor would it
be possible to produce any new Colour by Reflexions
and Refractions, and by consequence that the variety
of Colours depends upon the Composition of Light.

DEFINITION.

THE homogeneal Light and Rays which appear
red, or rather make Objects appear so, I call
Rubrifick or Red-making; those which make Objects
appear yellow, green, blue, and violet, I call Yellow-
making, Green-making, Blue-making, Violet-making,
and so of the rest. And if at any time I speak of Light
and Rays as coloured or endued with Colours, I
would be understood to speak not philosophically
and properly, but grossly, and accordingly to such
Conceptions as vulgar People in seeing all these Experiments would be apt to frame. For the Rays to
speak properly are not coloured. In them there is
nothing else than a certain Power and Disposition to

stir up a Sensation of this or that Colour. For as Sound in a Bell or musical String, or other sounding Body, is nothing but a trembling Motion, and in the Air nothing but that Motion propagated from the Object, and in the Sensorium 'tis a Sense of that Motion under the Form of Sound; so Colours in the Object are nothing but a Disposition to reflect this or that sort of Rays more copiously than the rest; in the Rays they are nothing but their Dispositions to propagate this or that Motion into the Sensorium, and in the Sensorium they are Sensations of those Motions under the Forms of Colours.

PROP. III. PROB. I.

To define the Refrangibility of the several sorts of homogeneal Light answering to the several Colours.

FOR determining this Problem I made the following Experiment.*

Exper. 7. When I had caused the Rectilinear Sides AF, GM, [in *Fig.* 4.] of the Spectrum of Colours made by the Prism to be distinctly defined, as in the fifth Experiment of the first Part of this Book is described, there were found in it all the homogeneal Colours in the same Order and Situation one among another as in the Spectrum of simple Light, described in the fourth Proposition of that Part. For the Circles of which the Spectrum of compound Light

* *See our* Author's Lect. Optic. *Part* II. *Sect.* II. *p.* 239.

PT is composed, and which in the middle Parts of the Spectrum interfere, and are intermix'd with one another, are not intermix'd in their outmost Parts where they touch those Rectilinear Sides AF and GM. And therefore, in those Rectilinear Sides when distinctly defined, there is no new Colour generated by Refraction. I observed also, that if any where between the two outmost Circles TMF and PGA a Right Line, as $\gamma\delta$, was cross to the Spectrum, so as both Ends to fall perpendicularly upon its Rectilinear Sides, there appeared one and the same Colour, and degree of Colour from one End of this Line to the other. I delineated therefore in a Paper the Perimeter of the Spectrum FAP GMT, and in trying the third Experiment of the first Part of this Book, I held the Paper so that the Spectrum might fall upon this delineated Figure, and agree with it exactly, whilst an Assistant, whose Eyes for distinguishing Colours were more critical than mine, did by Right Lines $\alpha\beta$, $\gamma\delta$, $\varepsilon\zeta$, &c. drawn cross the Spectrum, note the Confines of the Colours, that is of the red $M\alpha\beta F$, of the orange $\alpha\gamma\delta\beta$, of the yellow $\gamma\varepsilon\zeta\delta$, of the green $\varepsilon\eta\theta\zeta$, of the blue $\eta\iota\kappa\theta$, of the indico $\iota\lambda\mu\kappa$, and of the violet $\lambda GA\mu$. And this Operation being divers times repeated both in the same, and in several Papers, I found that the Observations agreed well enough with one another, and that the Rectilinear Sides MG and FA were by the said cross Lines divided after the manner of a Musical Chord. Let GM be produced to X, that MX may be equal to GM, and conceive GX, λX, ιX, ηX, εX, γX, αX, MX, to be in propor-

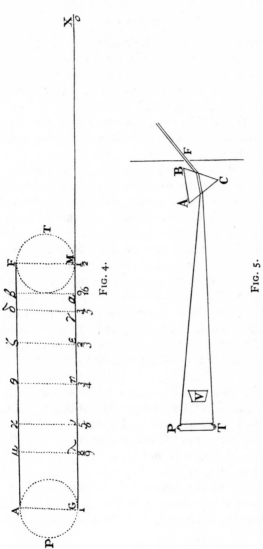

Fig. 4.

Fig. 5.

tion to one another, as the Numbers, $1, \frac{8}{9}, \frac{5}{6}, \frac{3}{4}, \frac{2}{3}, \frac{3}{5},$ $\frac{9}{16}, \frac{1}{2},$ and so to represent the Chords of the Key, and of a Tone, a third Minor, a fourth, a fifth, a sixth Major, a seventh and an eighth above that Key: And the Intervals Ma, $a\gamma$, $\gamma\varepsilon$, $\varepsilon\eta$, $\eta\iota$, $\iota\lambda$, and λG, will be the Spaces which the several Colours (red, orange, yellow, green, blue, indigo, violet) take up.

Now these Intervals or Spaces subtending the Differences of the Refractions of the Rays going to the Limits of those Colours, that is, to the Points M, a, γ, ε, η, ι, λ, G, may without any sensible Error be accounted proportional to the Differences of the Sines of Refraction of those Rays having one common Sine of Incidence, and therefore since the common Sine of Incidence of the most and least refrangible Rays out of Glass into Air was (by a Method described above) found in proportion to their Sines of Refraction, as 50 to 77 and 78, divide the Difference between the Sines of Refraction 77 and 78, as the Line GM is divided by those Intervals, and you will have $77, 77\frac{1}{8}, 77\frac{1}{5}, 77\frac{1}{3}, 77\frac{1}{2}, 77\frac{2}{3}, 77\frac{7}{9},$ 78, the Sines of Refraction of those Rays out of Glass into Air, their common Sine of Incidence being 50. So then the Sines of the Incidences of all the red-making Rays out of Glass into Air, were to the Sines of their Refractions, not greater than 50 to 77, nor less than 50 to $77\frac{1}{8}$, but they varied from one another according to all intermediate Proportions. And the Sines of the Incidences of the green-making Rays were to the Sines of their Refractions in all Proportions from that of 50 to $77\frac{1}{3}$, unto that of 50 to $77\frac{1}{2}$. And by the like Limits above-mentioned were the Refractions of the Rays belonging to the rest of the Colours defined, the Sines of the red-making Rays extending from 77 to $77\frac{1}{8}$, those of the orange-making from $77\frac{1}{8}$ to. $77\frac{1}{5}$, those of the yellow-making from $77\frac{1}{5}$ to $77\frac{1}{3}$, those of the green-making from $77\frac{1}{3}$ to $77\frac{1}{2}$, those of the blue-making from $77\frac{1}{2}$ to $77\frac{2}{3}$, those of the indigo-making from $77\frac{2}{3}$ to $77\frac{7}{9}$, and those of the violet from $77\frac{7}{9}$ to 78.

These are the Laws of the Refractions made out of Glass into Air, and thence by the third Axiom of the first Part of this Book, the Laws of the Refractions made out of Air into Glass are easily derived.

Exper. 8. I found moreover, that when Light goes out of Air through several contiguous refracting Mediums as through Water and Glass, and thence goes out again into Air, whether the refracting Superficies be parallel or inclin'd to one another, that Light as often as by contrary Refractions 'tis so corrected, that it emergeth in Lines parallel to those in which it was incident, continues ever after to be white. But if the emergent Rays be inclined to the incident, the Whiteness of the emerging Light will by degrees in passing on from the Place of Emergence, become tinged in its Edges with Colours. This I try'd by refracting Light with Prisms of Glass placed within a Prismatick Vessel of Water. Now those Colours argue a diverging and separation of the heterogeneous Rays from one another by means of their unequal Refractions, as in what follows will more fully appear. And, on the contrary, the permanent whiteness argues, that in like Incidences of the Rays there is no such separation of the emerging Rays, and by consequence no inequality of their whole Refractions. Whence I seem to gather the two following Theorems.

1. The Excesses of the Sines of Refraction of several sorts of Rays above their common Sine of Incidence when the Refractions are made out of divers denser Mediums immediately into one and the same rarer Medium, suppose of Air, are to one another in a given Proportion.

2. The Proportion of the Sine of Incidence to the Sine of Refraction of one and the same sort of Rays out of one Medium into another, is composed of the Proportion of the Sine of Incidence to the Sine of Refraction out of the first Medium into any third Medium, and of the Proportion of the Sine of Incidence to the Sine of Refraction out of that third Medium into the second Medium.

By the first Theorem the Refractions of the Rays of every sort made out of any Medium into Air are known by having the Refraction of the Rays of any one sort. As for instance, if the Refractions of the Rays of every sort out of Rain-water into Air be desired, let the common Sine of Incidence out of Glass into Air be subducted from the Sines of Refraction, and the Excesses will be 27, $27\frac{1}{8}$, $27\frac{1}{5}$, $27\frac{1}{3}$, $27\frac{1}{2}$, $27\frac{2}{3}$, $27\frac{7}{9}$, 28. Suppose now that the Sine of Incidence of the least refrangible Rays be to their Sine of Refraction out of Rain-water into Air as 3 to 4,

and say as 1 the difference of those Sines is to 3 the
Sine of Incidence, so is 27 the least of the Excesses
above-mentioned to a fourth Number 81; and 81
will be the common Sine of Incidence out of Rain-
water into Air, to which Sine if you add all the
abovementioned Excesses, you will have the desired
Sines of the Refractions 108, $108\frac{1}{8}$, $108\frac{1}{5}$, $108\frac{1}{3}$, $108\frac{1}{2}$,
$108\frac{2}{3}$, $108\frac{7}{9}$, 109.

By the latter Theorem the Refraction out of one
Medium into another is gathered as often as you
have the Refractions out of them both into any third
Medium. As if the Sine of Incidence of any Ray out
of Glass into Air be to its Sine of Refraction, as 20
to 31, and the Sine of Incidence of the same Ray out
of Air into Water, be to its Sine of Refraction as 4
to 3; the Sine of Incidence of that Ray out of Glass
into Water will be to its Sine of Refraction as 20 to
31 and 4 to 3 jointly, that is, as the Factum of 20 and
4 to the Factum of 31 and 3, or as 80 to 93.

And these Theorems being admitted into Opticks,
there would be scope enough of handling that
Science voluminously after a new manner,* not only
by teaching those things which tend to the perfection of Vision, but also by determining mathematically all kinds of Phænomena of Colours which could
be produced by Refractions. For to do this, there is
nothing else requisite than to find out the Separations of heterogeneous Rays, and their various
Mixtures and Proportions in every Mixture. By this

* *As is done in our* Author's Lect. Optic. *Part* I. *Sect.* III. *and*
IV. *and Part* II. *Sect.* II.

way of arguing I invented almost all the Phænomena
described in these Books, beside some others less
necessary to the Argument; and by the successes I
met with in the Trials, I dare promise, that to him
who shall argue truly, and then try all things with
good Glasses and sufficient Circumspection, the expected Event will not be wanting. But he is first to
know what Colours will arise from any others mix'd
in any assigned Proportion.

PROP. IV.   THEOR. III.

*Colours may be produced by Composition which shall
be like to the Colours of homogeneal Light as to
the Appearance of Colour, but not as to the Immutability of Colour and Constitution of Light.
And those Colours by how much they are more
compounded by so much are they less full and
intense, and by too much Composition they may be
diluted and weaken'd till they cease, and the
Mixture becomes white or grey. There may be also
Colours produced by Composition, which are not
fully like any of the Colours of homogeneal
Light.*

FOR a Mixture of homogeneal red and yellow
compounds an Orange, like in appearance of
Colour to that orange which in the series of unmixed
prismatick Colours lies between them; but the Light
of one orange is homogeneal as to Refrangibility,

and that of the other is heterogeneal, and the Colour
of the one, if viewed through a Prism, remains unchanged, that of the other is changed and resolved
into its component Colours red and yellow. And
after the same manner other neighbouring homogeneal Colours may compound new Colours, like the
intermediate homogeneal ones, as yellow and green,
the Colour between them both, and afterwards, if
blue be added, there will be made a green the middle
Colour of the three which enter the Composition.
For the yellow and blue on either hand, if they are
equal in quantity they draw the intermediate green
equally towards themselves in Composition, and so
keep it as it were in Æquilibrion, that it verge not
more to the yellow on the one hand, and to the blue
on the other, but by their mix'd Actions remain still
a middle Colour. To this mix'd green there may be
farther added some red and violet, and yet the green
will not presently cease, but only grow less full and
vivid, and by increasing the red and violet, it will
grow more and more dilute, until by the prevalence
of the added Colours it be overcome and turned into
whiteness, or some other Colour. So if to the Colour
of any homogeneal Light, the Sun's white Light composed of all sorts of Rays be added, that Colour will
not vanish or change its Species, but be diluted, and
by adding more and more white it will be diluted
more and more perpetually. Lastly, If red and violet
be mingled, there will be generated according to
their various Proportions various Purples, such as
are not like in appearance to the Colour of any

homogeneal Light, and of these Purples mix'd with
yellow and blue may be made other new Colours.

PROP. V.   THEOR. IV.

Whiteness and all grey Colours between white and
black, may be compounded of Colours, and the
whiteness of the Sun's Light is compounded of all
the primary Colours mix'd in a due Proportion.

The PROOF by Experiments.

Exper. 9. THE Sun shining into a dark Chamber through a little round hole in the
Window-shut, and his Light being there refracted by
a Prism to cast his coloured Image PT [in *Fig.* 5.]
upon the opposite Wall: I held a white Paper V to
that image in such manner that it might be illuminated by the colour'd Light reflected from thence,
and yet not intercept any part of that Light in its
passage from the Prism to the Spectrum. And I
found that when the Paper was held nearer to any
Colour than to the rest, it appeared of that Colour to
which it approached nearest; but when it was equally
or almost equally distant from all the Colours, so
that it might be equally illuminated by them all it
appeared white. And in this last situation of the
Paper, if some Colours were intercepted, the Paper
lost its white Colour, and appeared of the Colour of
the rest of the Light which was not intercepted. So
then the Paper was illuminated with Lights of various

Colours, namely, red, yellow, green, blue and violet, and every part of the Light retained its proper Colour, until it was incident on the Paper, and became reflected thence to the Eye; so that if it had been either alone (the rest of the Light being intercepted) or if it had abounded most, and been predominant in the Light reflected from the Paper, it would have tinged the Paper with its own Colour; and yet being mixed with the rest of the Colours in a due proportion, it made the Paper look white, and therefore by a Composition with the rest produced that Colour. The several parts of the coloured Light reflected from the Spectrum, whilst they are propagated from thence through the Air, do perpetually retain their proper Colours, because wherever they fall upon the Eyes of any Spectator, they make the several parts of the Spectrum to appear under their proper Colours. They retain therefore their proper Colours when they fall upon the Paper V, and so by the confusion and perfect mixture of those Colours compound the whiteness of the Light reflected from thence.

Exper. 10. Let that Spectrum or solar Image PT [in *Fig.* 6.] fall now upon the Lens MN above four Inches broad, and about six Feet distant from the Prism ABC and so figured that it may cause the coloured Light which divergeth from the Prism to converge and meet again at its Focus G, about six or eight Feet distant from the Lens, and there to fall perpendicularly upon a white Paper DE. And if you move this Paper to and fro, you will perceive that near the Lens, as at *de*, the whole solar Image (suppose at *pt*) will appear upon it intensely coloured after the manner above-explained, and that by receding from the Lens those Colours will perpetually come towards one another, and by mixing more and more dilute one another continually, until at length the Paper come to the Focus G, where by a perfect mixture they will wholly vanish and be converted into whiteness, the whole Light appearing now upon the Paper like a little white Circle. And afterwards by receding farther from the Lens, the Rays which before converged will now cross one another in the Focus G, and diverge from thence, and thereby make the Colours to appear again, but yet in a contrary order; suppose at $\delta\varepsilon$, where the red *t* is now above which before was below, and the violet *p* is below which before was above.

Let us now stop the Paper at the Focus G, where the Light appears totally white and circular, and let us consider its whiteness. I say, that this is composed of the converging Colours. For if any of those Colours be intercepted at the Lens, the whiteness will cease and degenerate into that Colour which ariseth from the composition of the other Colours which are not intercepted. And then if the intercepted Colours be let pass and fall upon that compound Colour, they mix with it, and by their mixture restore the whiteness. So if the violet, blue and green be intercepted, the remaining yellow, orange and red will compound upon the Paper an orange, and then if the intercepted Colours be let pass, they will fall

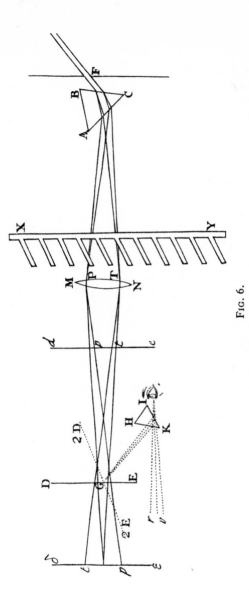

FIG. 6.

upon this compounded orange, and together with it decompound a white. So also if the red and violet be intercepted, the remaining yellow, green and blue, will compound a green upon the Paper, and then the red and violet being let pass will fall upon this green, and together with it decompound a white. And that in this Composition of white the several Rays do not suffer any Change in their colorific Qualities by acting upon one another, but are only mixed, and by a mixture of their Colours produce white, may farther appear by these Arguments.

If the Paper be placed beyond the Focus G, suppose at $\delta\varepsilon$, and then the red Colour at the Lens be alternately intercepted, and let pass again, the violet Colour on the Paper will not suffer any Change thereby, as it ought to do if the several sorts of Rays acted upon one another in the Focus G, where they cross. Neither will the red upon the Paper be changed by any alternate stopping, and letting pass the violet which crosseth it.

And if the Paper be placed at the Focus G, and the white round Image at G be viewed through the Prism HIK, and by the Refraction of that Prism be translated to the place rv, and there appear tinged with various Colours, namely, the violet at v and red at r, and others between, and then the red Colours at the Lens be often stopp'd and let pass by turns, the red at r will accordingly disappear, and return as often, but the violet at v will not thereby suffer any Change. And so by stopping and letting pass alternately the blue at the Lens, the blue at v will accord-

ingly disappear and return, without any Change
made in the red at r. The red therefore depends on
one sort of Rays, and the blue on another sort, which
in the Focus G where they are commix'd, do not act
on one another. And there is the same Reason of the
other Colours.

I considered farther, that when the most refrangible Rays Pp, and the least refrangible ones Tt, are
by converging inclined to one another, the Paper, if
held very oblique to those Rays in the Focus G,
might reflect one sort of them more copiously than
the other sort, and by that Means the reflected Light
would be tinged in that Focus with the Colour of the
predominant Rays, provided those Rays severally retained their Colours, or colorific Qualities in the
Composition of White made by them in that Focus.
But if they did not retain them in that White, but
became all of them severally endued there with a
Disposition to strike the Sense with the Perception of
White, then they could never lose their Whiteness by
such Reflexions. I inclined therefore the Paper to the
Rays very obliquely, as in the second Experiment of
this second Part of the first Book, that the most refrangible Rays, might be more copiously reflected
than the rest, and the Whiteness at Length changed
successively into blue, indigo, and violet. Then I inclined it the contrary Way, that the least refrangible
Rays might be more copious in the reflected Light
than the rest, and the Whiteness turned successively
to yellow, orange, and red.

Lastly, I made an Instrument XY in fashion of a

Comb, whose Teeth being in number sixteen, were about an Inch and a half broad, and the Intervals of the Teeth about two Inches wide. Then by interposing successively the Teeth of this Instrument near the Lens, I intercepted Part of the Colours by the interposed Tooth, whilst the rest of them went on through the Interval of the Teeth to the Paper DE, and there painted a round Solar Image. But the Paper I had first placed so, that the Image might appear white as often as the Comb was taken away ; and then the Comb being as was said interposed, that Whiteness by reason of the intercepted Part of the Colours at the Lens did always change into the Colour compounded of those Colours which were not intercepted, and that Colour was by the Motion of the Comb perpetually varied so, that in the passing of every Tooth over the Lens all these Colours, red, yellow, green, blue, and purple, did always succeed one another. I caused therefore all the Teeth to pass successively over the Lens, and when the Motion was slow, there appeaɪed a perpetual Succession of the Colours upon the Paper: But if I so much accelerated the Motion, that the Colours by reason of their quick Succession could not be distinguished from one another, the Appearance of the single Colours ceased. There was no red, no yellow, no green, no blue, nor purple to be seen any longer, but from a Confusion of them all there arose one uniform white Colour. Of the Light which now by the Mixture of all the Colours appeared white, there was no Part really white. One Part was red, another yellow, a third green, a fourth blue, a fifth purple, and every Part retains its proper Colour till it strike the Sensorium. If the Impressions follow one another slowly, so that they may be severally perceived, there is made a distinct Sensation of all the Colours one after another in a continual Succession. But if the Impressions follow one another so quickly, that they cannot be severally perceived, there ariseth out of them all one common Sensation, which is neither of this Colour alone nor of that alone, but hath it self indifferently to 'em all, and this is a Sensation of Whiteness. By the Quickness of the Successions, the Impressions of the several Colours are confounded in the Sensorium, and out of that Confusion ariseth a mix'd Sensation. If a burning Coal be nimbly moved round in a Circle with Gyrations continually repeated, the whole Circle will appear like Fire; the reason of which is, that the Sensation of the Coal in the several Places of that Circle remains impress'd on the Sensorium, until the Coal return again to the same Place. And so in a quick Consecution of the Colours the Impression of every Colour remains in the Sensorium, until a Revolution of all the Colours be compleated, and that first Colour return again. The Impressions therefore of all the successive Colours are at once in the Sensorium, and jointly stir up a Sensation of them all; and so it is manifest by this Experiment, that the commix'd Impressions of all the Colours do stir up and beget a Sensation of white, that is, that Whiteness is compounded of all the Colours.

And if the Comb be now taken away, that all the Colours may at once pass from the Lens to the Paper, and be there intermixed, and together reflected thence to the Spectator's Eyes; their Impressions on the Sensorium being now more subtilly and perfectly commixed there, ought much more to stir up a Sensation of Whiteness.

You may instead of the Lens use two Prisms HIK and LMN, which by refracting the coloured Light the contrary Way to that of the first Refraction, may make the diverging Rays converge and meet again in G, as you see represented in the seventh Figure. For where they meet and mix, they will compose a white Light, as when a Lens is used.

Expt. 11. Let the Sun's coloured Image PT [in *Fig.* 8.] fall upon the Wall of a dark Chamber, as in the third Experiment of the first Book, and let the same be viewed through a Prism *abc*, held parallel to the Prism ABC, by whose Refraction that Image was made, and let it now appear lower than before, suppose in the Place S over-against the red Colour T. And if you go near to the Image PT, the Spectrum S will appear oblong and coloured like the Image PT; but if you recede from it, the Colours of the spectrum S will be contracted more and more, and at length vanish, that Spectrum S becoming perfectly round and white; and if you recede yet farther, the Colours will emerge again, but in a contrary Order. Now that Spectrum S appears white in that Case, when the Rays of several sorts which converge from the several Parts of the Image PT, to the Prism *abc*,

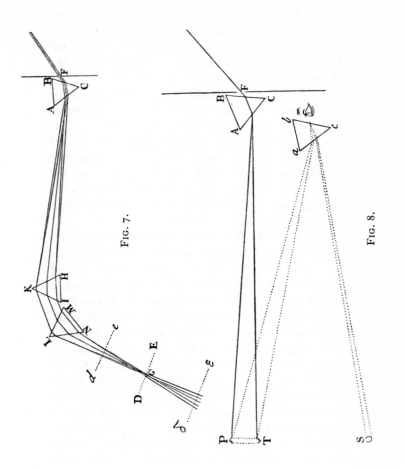

FIG. 7.

FIG. 8.

are so refracted unequally by it, that in their Passage from the Prism to the Eye they may diverge from one and the same Point of the Spectrum S, and so fall afterwards upon one and the same Point in the bottom of the Eye, and there be mingled.

And farther, if the Comb be here made use of, by whose Teeth the Colours at the Image PT may be successively intercepted; the Spectrum S, when the Comb is moved slowly, will be perpetually tinged with successive Colours: But when by accelerating the Motion of the Comb, the Succession of the Colours is so quick that they cannot be severally seen, that Spectrum S, by a confused and mix'd Sensation of them all, will appear white.

Exper. 12. The Sun shining through a large Prism ABC [in *Fig.* 9.] upon a Comb XY, placed immediately behind the Prism, his Light which passed through the Interstices of the Teeth fell upon a white Paper DE. The Breadths of the Teeth were equal to their Interstices, and seven Teeth together with their Interstices took up an Inch in Breadth. Now, when the Paper was about two or three Inches distant from the Comb, the Light which passed through its several Interstices painted so many Ranges of Colours, *kl*, *mn*, *op*, *qr*, &c. which were parallel to one another, and contiguous, and without any Mixture of white. And these Ranges of Colours, if the Comb was moved continually up and down with a reciprocal Motion, ascended and descended in the Paper, and when the Motion of the Comb was so quick, that the Colours could not be distinguished

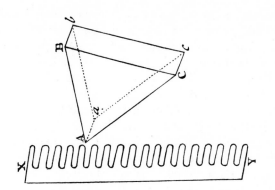

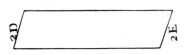

Fig. 9.

from one another, the whole Paper by their Confusion and Mixture in the Sensorium appeared white.

Let the Comb now rest, and let the Paper be removed farther from the Prism, and the several Ranges
of Colours will be dilated and expanded into one
another more and more, and by mixing their Colours
will dilute one another, and at length, when the
distance of the Paper from the Comb is about a Foot,
or a little more (suppose in the Place 2D 2E) they
will so far dilute one another, as to become white.

With any Obstacle, let all the Light be now
stopp'd which passes through any one Interval of
the Teeth, so that the Range of Colours which comes
from thence may be taken away, and you will see the
Light of the rest of the Ranges to be expanded into
the Place of the Range taken away, and there to be
coloured. Let the intercepted Range pass on as before, and its Colours falling upon the Colours of the
other Ranges, and mixing with them, will restore the
Whiteness.

Let the Paper 2D 2E be now very much inclined
to the Rays, so that the most refrangible Rays may be
more copiously reflected than the rest, and the white
Colour of the Paper through the Excess of those
Rays will be changed into blue and violet. Let the
Paper be as much inclined the contrary way, that the
least refrangible Rays may be now more copiously
reflected than the rest, and by their Excess the Whiteness will be changed into yellow and red. The several
Rays therefore in that white Light do retain their
colorific Qualities, by which those of any sort, when-

ever they become more copious than the rest, do by
their Excess and Predominance cause their proper
Colour to appear.

And by the same way of arguing, applied to the
third Experiment of this second Part of the first Book,
it may be concluded, that the white Colour of all

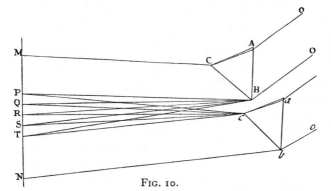

FIG. 10.

refracted Light at its very first Emergence, where it
appears as white as before its Incidence, is compounded of various Colours.

Exper. 13. In the foregoing Experiment the several
Intervals of the Teeth of the Comb do the Office of
so many Prisms, every Interval producing the Phænomenon of one Prism. Whence instead of those
Intervals using several Prisms, I try'd to compound
Whiteness by mixing their Colours, and did it by
using only three Prisms, as also by using only two as
follows. Let two Prisms ABC and *abc*, [in *Fig.* 10.]
whose refracting Angles B and *b* are equal, be so

placed parallel to one another, that the refracting
Angle B of the one may touch the Angle c at the Base
of the other, and their Planes CB and cb, at which the
Rays emerge, may lie in Directum. Then let the
Light trajected through them fall upon the Paper
MN, distant about 8 or 12 Inches from the Prisms.
And the Colours generated by the interior Limits B
and c of the two Prisms, will be mingled at PT, and
there compound white. For if either Prism be taken
away, the Colours made by the other will appear in
that Place PT, and when the Prism is restored to its
Place again, so that its Colours may there fall upon
the Colours of the other, the Mixture of them both
will restore the Whiteness.

This Experiment succeeds also, as I have tried,
when the Angle b of the lower Prism, is a little greater
than the Angle B of the upper, and between the interior Angles B and c, there intercedes some Space
Bc, as is represented in the Figure, and the refracting
Planes BC and bc, are neither in Directum, nor parallel to one another. For there is nothing more requisite to the Success of this Experiment, than that
the Rays of all sorts may be uniformly mixed upon
the Paper in the Place PT. If the most refrangible
Rays coming from the superior Prism take up all the
Space from M to P, the Rays of the same sort which
come from the inferior Prism ought to begin at P,
and take up all the rest of the Space from thence towards N. If the least refrangible Rays coming from
the superior Prism take up the Space MT, the Rays
of the same kind which come from the other Prism

ought to begin at T, and take up the remaining Space
TN. If one sort of the Rays which have intermediate
Degrees of Refrangibility, and come from the superior Prism be extended through the Space MQ, and
another sort of those Rays through the Space MR,
and a third sort of them through the Space MS, the
same sorts of Rays coming from the lower Prism,
ought to illuminate the remaining Spaces QN, RN,
SN, respectively. And the same is to be understood
of all the other sorts of Rays. For thus the Rays of
every sort will be scattered uniformly and evenly
through the whole Space MN, and so being every
where mix'd in the same Proportion, they must
every where produce the same Colour. And therefore, since by this Mixture they produce white in the
Exterior Spaces MP and TN, they must also produce
white in the Interior Space PT. This is the reason
of the Composition by which Whiteness was produced in this Experiment, and by what other way
soever I made the like Composition, the Result was
Whiteness.

Lastly, If with the Teeth of a Comb of a due Size,
the coloured Lights of the two Prisms which fall upon
the Space PT be alternately intercepted, that Space
PT, when the Motion of the Comb is slow, will
always appear coloured, but by accelerating the
Motion of the Comb so much that the successive
Colours cannot be distinguished from one another,
it will appear white.

Exper. 14. Hitherto I have produced Whiteness
by mixing the Colours of Prisms. If now the Colours

of natural Bodies are to be mingled, let Water a little thicken'd with Soap be agitated to raise a Froth, and after that Froth has stood a little, there will appear to one that shall view it intently various Colours every where in the Surfaces of the several Bubbles; but to one that shall go so far off, that he cannot distinguish the Colours from one another, the whole Froth will grow white with a perfect Whiteness.

Exper. 15. Lastly, In attempting to compound a white, by mixing the coloured Powders which Painters use, I consider'd that all colour'd Powders do suppress and stop in them a very considerable Part of the Light by which they are illuminated. For they become colour'd by reflecting the Light of their own Colours more copiously, and that of all other Colours more sparingly, and yet they do not reflect the Light of their own Colours so copiously as white Bodies do. If red Lead, for instance, and a white Paper, be placed in the red Light of the colour'd Spectrum made in a dark Chamber by the Refraction of a Prism, as is described in the third Experiment of the first Part of this Book; the Paper will appear more lucid than the red Lead, and therefore reflects the red-making Rays more copiously than red Lead doth. And if they be held in the Light of any other Colour, the Light reflected by the Paper will exceed the Light reflected by the red Lead in a much greater Proportion. And the like happens in Powders of other Colours. And therefore by mixing such Powders, we are not to expect a strong and full

White, such as is that of Paper, but some dusky obscure one, such as might arise from a Mixture of Light and Darkness, or from white and black, that is, a grey, or dun, or russet brown, such as are the Colours of a Man's Nail, of a Mouse, of Ashes, of ordinary Stones, of Mortar, of Dust and Dirt in High-ways, and the like. And such a dark white I have often produced by mixing colour'd Powders. For thus one Part of red Lead, and five Parts of *Viride Æris*, composed a dun Colour like that of a Mouse. For these two Colours were severally so compounded of others, that in both together were a Mixture of all Colours; and there was less red Lead used than *Viride Æris*, because of the Fulness of its Colour. Again, one Part of red Lead, and four Parts of blue Bise, composed a dun Colour verging a little to purple, and by adding to this a certain Mixture of Orpiment and *Viride Æris* in a due Proportion, the Mixture lost its purple Tincture, and became perfectly dun. But the Experiment succeeded best without Minium thus. To Orpiment I added by little and little a certain full bright purple, which Painters use, until the Orpiment ceased to be yellow, and became of a pale red. Then I diluted that red by adding a little *Viride Æris*, and a little more blue Bise than *Viride Æris*, until it became of such a grey or pale white, as verged to no one of the Colours more than to another. For thus it became of a Colour equal in Whiteness to that of Ashes, or of Wood newly cut, or of a Man's Skin. The Orpiment reflected more Light than did any other of the Powders, and there-

fore conduced more to the Whiteness of the compounded Colour than they. To assign the Proportions
accurately may be difficult, by reason of the different
Goodness of Powders of the same kind. Accordingly,
as the Colour of any Powder is more or less full and
luminous, it ought to be used in a less or greater
Proportion.

Now, considering that these grey and dun Colours may be also produced by mixing Whites and
Blacks, and by consequence differ from perfect
Whites, not in Species of Colours, but only in degree
of Luminousness, it is manifest that there is nothing
more requisite to make them perfectly white than to
increase their Light sufficiently; and, on the contrary,
if by increasing their Light they can be brought to
perfect Whiteness, it will thence also follow, that
they are of the same Species of Colour with the best
Whites, and differ from them only in the Quantity of
Light. And this I tried as follows. I took the third of
the above-mention'd grey Mixtures, (that which was
compounded of Orpiment, Purple, Bise, and *Viride
Æris*) and rubbed it thickly upon the Floor of my
Chamber, where the Sun shone upon it through the
opened Casement; and by it, in the shadow, I laid a
Piece of white Paper of the same Bigness. Then going
from them to the distance of 12 or 18 Feet, so that
I could not discern the Unevenness of the Surface
of the Powder, nor the little Shadows let fall from
the gritty Particles thereof; the Powder appeared intensely white, so as to transcend even the Paper it
self in Whiteness, especially if the Paper were a little

shaded from the Light of the Clouds, and then the
Paper compared with the Powder appeared of such a
grey Colour as the Powder had done before. But by
laying the Paper where the Sun shines through the
Glass of the Window, or by shutting the Window
that the Sun might shine through the Glass upon the
Powder, and by such other fit Means of increasing
or decreasing the Lights wherewith the Powder and
Paper were illuminated, the Light wherewith the
Powder is illuminated may be made stronger in such
a due Proportion than the Light wherewith the
Paper is illuminated, that they shall both appear
exactly alike in Whiteness. For when I was trying
this, a Friend coming to visit me, I stopp'd him at
the Door, and before I told him what the Colours
were, or what I was doing; I asked him, Which of
the two Whites were the best, and wherein they differed? And after he had at that distance viewed them
well, he answer'd, that they were both good Whites,
and that he could not say which was best, nor wherein their Colours differed. Now, if you consider, that
this White of the Powder in the Sun-shine was compounded of the Colours which the component Powders (Orpiment, Purple, Bise, and *Viride Æris*) have
in the same Sun-shine, you must acknowledge by
this Experiment, as well as by the former, that perfect Whiteness may be compounded of Colours.

From what has been said it is also evident, that the
Whiteness of the Sun's Light is compounded of all
the Colours wherewith the several sorts of Rays
whereof that Light consists, when by their several

Refrangibilities they are separated from one another, do tinge Paper or any other white Body whereon they fall. For those Colours (by *Prop.* II. *Part 2.*) are unchangeable, and whenever all those Rays with those their Colours are mix'd again, they reproduce the same white Light as before.

PROP. VI. Prob. II.

In a mixture of Primary Colours, the Quantity and Quality of each being given, to know the Colour of the Compound.

WITH the Center O [in *Fig.* 11.] and Radius OD describe a Circle ADF, and distinguish its Circumference into seven Parts DE, EF, FG, GA, AB, BC, CD, proportional to the seven Musical Tones or Intervals of the eight Sounds, *Sol, la, fa, sol, la, mi, fa, sol*, contained in an eight, that is, proportional to the Number $\frac{1}{9}$, $\frac{1}{16}$, $\frac{1}{10}$, $\frac{1}{9}$, $\frac{1}{16}$, $\frac{1}{16}$, $\frac{1}{9}$. Let the first Part DE represent a red Colour, the second EF orange, the third FG yellow, the fourth CA green, the fifth AB blue, the sixth BC indigo, and the seventh CD violet. And conceive that these are all the Colours of uncompounded Light gradually passing into one another, as they do when made by Prisms; the Circumference DEFGABCD, representing the whole Series of Colours from one end of the Sun's colour'd Image to the other, so that from D to E be all degrees of red, at E the mean Colour

between red and orange, from E to F all degrees of
orange, at F the mean between orange and yellow,
from F to G all degrees of yellow, and so on. Let p
be the Center of Gravity of the Arch DE, and $q, r, s,$
t, u, x, the Centers of Gravity of the Arches EF, FG,
GA, AB, BC, and CD respectively, and about those

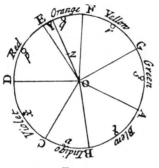

FIG. 11.

Centers of Gravity let Circles proportional to the
Number of Rays of each Colour in the given Mixture
be describ'd: that is, the Circle p proportional to the
Number of the red-making Rays in the Mixture, the
Circle q proportional to the Number of the orange-
making Rays in the Mixture, and so of the rest. Find
the common Center of Gravity of all those Circles,
p, q, r, s, t, u, x. Let that Center be Z; and from the
Center of the Circle ADF, through Z to the Circum-
ference, drawing the Right Line OY, the Place of
the Point Y in the Circumference shall shew the
Colour arising from the Composition of all the
Colours in the given Mixture, and the Line OZ shall

be proportional to the Fulness or Intenseness of the
Colour, that is, to its distance from Whiteness. As if
Y fall in the middle between F and G, the compounded Colour shall be the best yellow; if Y verge
from the middle towards F or G, the compound
Colour shall accordingly be a yellow, verging towards orange or green. If Z fall upon the Circumference, the Colour shall be intense and florid in the
highest Degree; if it fall in the mid-way between the
Circumference and Center, it shall be but half so
intense, that is, it shall be such a Colour as would be
made by diluting the intensest yellow with an equal
quantity of whiteness; and if it fall upon the center O,
the Colour shall have lost all its intenseness, and become a white. But it is to be noted, That if the point
Z fall in or near the line OD, the main ingredients
being the red and violet, the Colour compounded
shall not be any of the prismatick Colours, but a
purple, inclining to red or violet, accordingly as the
point Z lieth on the side of the line DO towards E
or towards C, and in general the compounded violet
is more bright and more fiery than the uncompounded. Also if only two of the primary Colours
which in the circle are opposite to one another be
mixed in an equal proportion, the point Z shall fall
upon the center O, and yet the Colour compounded
of those two shall not be perfectly white, but some
faint anonymous Colour. For I could never yet by
mixing only two primary Colours produce a perfect
white. Whether it may be compounded of a mixture
of three taken at equal distances in the circumference

I do not know, but of four or five I do not much question but it may. But these are Curiosities of little or no moment to the understanding the Phænomena of Nature. For in all whites produced by Nature, there uses to be a mixture of all sorts of Rays, and by consequence a composition of all Colours.

To give an instance of this Rule; suppose a Colour is compounded of these homogeneal Colours, of violet one part, of indigo one part, of blue two parts, of green three parts, of yellow five parts, of orange six parts, and of red ten parts. Proportional to these parts describe the Circles x, v, t, s, r, q, p, respectively, that is, so that if the Circle x be one, the Circle v may be one, the Circle t two, the Circle s three, and the Circles r, q and p, five, six and ten. Then I find Z the common center of gravity of these Circles, and through Z drawing the Line OY, the Point Y falls upon the circumference between E and F, something nearer to E than to F, and thence I conclude, that the Colour compounded of these Ingredients will be an orange, verging a little more to red than to yellow. Also I find that OZ is a little less than one half of OY, and thence I conclude, that this orange hath a little less than half the fulness or intenseness of an uncompounded orange; that is to say, that it is such an orange as may be made by mixing an homogeneal orange with a good white in the proportion of the Line OZ to the Line ZY, this Proportion being not of the quantities of mixed orange and white Powders, but of the quantities of the Lights reflected from them.

This Rule I conceive accurate enough for practice, though not mathematically accurate; and the truth of it may be sufficiently proved to Sense, by stopping any of the Colours at the Lens in the tenth Experiment of this Book. For the rest of the Colours which are not stopp'd, but pass on to the Focus of the Lens, will there compound either accurately or very nearly such a Colour, as by this Rule ought to result from their Mixture.

PROP. VII. Theor. V.

All the Colours in the Universe which are made by Light, and depend not on the Power of Imagination, are either the Colours of homogeneal Lights, or compounded of these, and that either accurately or very nearly, according to the Rule of the foregoing Problem.

FOR it has been proved (in *Prop.* 1. *Part* 2.) that the changes of Colours made by Refractions do not arise from any new Modifications of the Rays impress'd by those Refractions, and by the various Terminations of Light and Shadow, as has been the constant and general Opinion of Philosophers. It has also been proved that the several Colours of the homogeneal Rays do constantly answer to their degrees of Refrangibility, (*Prop.* 1. *Part* 1. and *Prop.* 2. *Part* 2.) and that their degrees of Refrangibility cannot be changed by Refractions and Reflexions (*Prop.*

2. *Part* 1.) and by consequence that those their Colours are likewise immutable. It has also been proved directly by refracting and reflecting homogeneal Lights apart, that their Colours cannot be changed, (*Prop.* 2. *Part* 2.) It has been proved also, that when the several sorts of Rays are mixed, and in crossing pass through the same space, they do not act on one another so as to change each others colorific qualities. (*Exper.* 10. *Part* 2.) but by mixing their Actions in the Sensorium beget a Sensation differing from what either would do apart, that is a Sensation of a mean Colour between their proper Colours; and particularly when by the concourse and mixtures of all sorts of Rays, a white Colour is produced, the white is a mixture of all the Colours which the Rays would have apart, (*Prop.* 5. *Part* 2) The Rays in that mixture do not lose or alter their several colorific qualities, but by all their various kinds of Actions mix'd in the Sensorium, beget a Sensation of a middling Colour between all their Colours, which is whiteness. For whiteness is a mean between all Colours, having it self indifferently to them all, so as with equal facility to be tinged with any of them. A red Powder mixed with a little blue, or a blue with a little red, doth not presently lose its Colour, but a white Powder mix'd with any Colour is presently tinged with that Colour, and is equally capable of being tinged with any Colour whatever. It has been shewed also, that as the Sun's Light is mix'd of all sorts of Rays, so its whiteness is a mixture of the Colours of all sorts of Rays; those Rays having from the beginning their

several colorific qualities as well as their several Refrangibilities, and retaining them perpetually unchanged notwithstanding any Refractions or Reflexions they may at any time suffer, and that whenever
any sort of the Sun's Rays is by any means (as by
Reflexion in *Exper.* 9, and 10. *Part* 1. or by Refraction as happens in all Refractions) separated from the
rest, they then manifest their proper Colours. These
things have been prov'd, and the sum of all this
amounts to the Proposition here to be proved. For if
the Sun's Light is mix'd of several sorts of Rays,
each of which have originally their several Refrangibilities and colorific Qualities, and notwithstanding
their Refractions and Reflexions, and their various
Separations or Mixtures, keep those their original
Properties perpetually the same without alteration;
then all the Colours in the World must be such as
constantly ought to arise from the original colorific
qualities of the Rays whereof the Lights consist by
which those Colours are seen. And therefore if the
reason of any Colour whatever be required, we have
nothing else to do than to consider how the Rays in
the Sun's Light have by Reflexions or Refractions,
or other causes, been parted from one another, or
mixed together; or otherwise to find out what sorts of
Rays are in the Light by which that Colour is made,
and in what Proportion; and then by the last Problem
to learn the Colour which ought to arise by mixing
those Rays (or their Colours) in that proportion. I
speak here of Colours so far as they arise from Light.
For they appear sometimes by other Causes, as when

by the power of Phantasy we see Colours in a Dream,
or a Mad-man sees things before him which are not
there; or when we see Fire by striking the Eye, or see
Colours like the Eye of a Peacock's Feather, by pressing our Eyes in either corner whilst we look the other
way. Where these and such like Causes interpose
not, the Colour always answers to the sort or sorts
of the Rays whereof the Light consists, as I have constantly found in whatever Phænomena of Colours I
have hitherto been able to examine. I shall in the
following Propositions give instances of this in the
Phænomena of chiefest note.

PROP. VIII. Prob. III.

*By the discovered Properties of Light to explain the
Colours made by Prisms.*

LET ABC [in *Fig.* 12.] represent a Prism refracting the Light of the Sun, which comes into a
dark Chamber through a hole Fφ almost as broad as
the Prism, and let MN represent a white Paper on
which the refracted Light is cast, and suppose the
most refrangible or deepest violet-making Rays fall
upon the Space Pπ, the least refrangible or deepest
red-making Rays upon the Space Tτ, the middle sort
between the indigo-making and blue-making Rays
upon the Space Qχ, the middle sort of the green-
making Rays upon the Space R, the middle sort
between the yellow-making and orange-making Rays

upon the Space Sσ, and other intermediate sorts upon intermediate Spaces. For so the Spaces upon which the several sorts adequately fall will by reason of the different Refrangibility of those sorts be one lower than another. Now if the Paper MN be so near the Prism that the Spaces PT and $\pi\tau$ do not interfere with one another, the distance between them Tπ will be illuminated by all the sorts of Rays in that proportion to one another which they have at their very first coming out of the Prism, and consequently be white. But the Spaces PT and $\pi\tau$ on either hand, will not be illuminated by them all, and therefore will appear coloured. And particularly at P, where the outmost violet-making Rays fall alone, the Colour must be the deepest violet. At Q where the violet-making and indigo-making Rays are mixed, it must be a violet inclining much to indigo. At R where the violet-making, indigo-making, blue-making, and one half of the green-making Rays are mixed, their Colours must (by the construction of the second Problem) compound a middle Colour between indigo and blue. At S where all the Rays are mixed, except the red-making and orange-making, their Colours ought by the same Rule to compound a faint blue, verging more to green than indigo. And in the progress from S to T, this blue will grow more and more faint and dilute, till at T, where all the Colours begin to be mixed, it ends in whiteness.

So again, on the other side of the white at τ, where the least refrangible or utmost red-making Rays are alone, the Colour must be the deepest red. At σ the

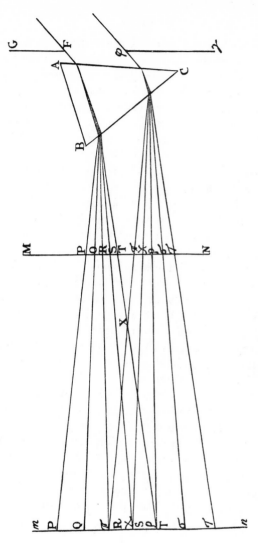

FIG. 12.

mixture of red and orange will compound a red inclining to orange. At ϱ the mixture of red, orange, yellow, and one half of the green must compound a middle Colour between orange and yellow. At χ the mixture of all Colours but violet and indigo will compound a faint yellow, verging more to green than to orange. And this yellow will grow more faint and dilute continually in its progress from χ to π, where by a mixture of all sorts of Rays it will become white.

These Colours ought to appear were the Sun's Light perfectly white: But because it inclines to yellow, the Excess of the yellow-making Rays whereby 'tis tinged with that Colour, being mixed with the faint blue between S and T, will draw it to a faint green. And so the Colours in order from P to τ ought to be violet, indigo, blue, very faint green, white, faint yellow, orange, red. Thus it is by the computation: And they that please to view the Colours made by a Prism will find it so in Nature.

These are the Colours on both sides the white when the Paper is held between the Prism and the Point X where the Colours meet, and the interjacent white vanishes. For if the Paper be held still farther off from the Prism, the most refrangible and least refrangible Rays will be wanting in the middle of the Light, and the rest of the Rays which are found there, will by mixture produce a fuller green than before. Also the yellow and blue will now become less compounded, and by consequence more intense than before. And this also agrees with experience.

REPLICA OF SIR ISAAC NEWTON'S
ORIGINAL REFLECTING TELESCOPE (1671)

Dʳ Halley

Orbells building in Kensington.
March 7ᵗʰ, 172⅞.

I thank you for the Table you sent me of the motion of the Comet of 1680 in a Parabolic Orb so as to answer to Kirks Observations as well as to Flamsteds. It answers all their Observations well enough for my purpose. But you have omitted the distances of the Comet from the Sun in parts of the mean distance of the earth from the Sun divided into 100000 equal parts: such parts as the Latus rectum of this Parabolic Orb consists of 2508. These Distances you have computed already in your papers in wᶜʰ you calculated this Table, & you need only to copy them from thence. I have inclosed a copy of your Table with a vacant column for these distances, & beg the favour of you to fill it up by inserting these distances out of those your loose papers in wᶜʰ you made your calculations of this Table. The distances are inserted in your Table published in the second edition of my Principles pag 459. I intend still to keep that Table & add this new one to it if you please to fill up the column of distances in the same manner that the two Tables may be like one another. And by the help of this new Table I shall be able to make the schemes of the motion of this Comet more perfect. I am

Yoᵉ humble servant.

Isaac. Newton.

NEWTON REFERS TO THE COMET WHOSE ORBIT
HALLEY HAD CALCULATED BY APPLYING NEWTON'S
PRINCIPLES OF PLANETARY MOTION.

And if one look through a Prism upon a white Object encompassed with blackness or darkness, the reason of the Colours arising on the edges is much the same, as will appear to one that shall a little consider it. If a black Object be encompassed with a white one, the Colours which appear through the Prism are to be derived from the Light of the white one, spreading into the Regions of the black, and therefore they appear in a contrary order to that, when a white Object is surrounded with black. And the same is to be understood when an Object is viewed, whose parts are some of them less luminous than others. For in the borders of the more and less luminous Parts, Colours ought always by the same Principles to arise from the Excess of the Light of the more luminous, and to be of the same kind as if the darker parts were black, but yet to be more faint and dilute.

What is said of Colours made by Prisms may be easily applied to Colours made by the Glasses of Telescopes or Microscopes, or by the Humours of the Eye. For if the Object-glass of a Telescope be thicker on one side than on the other, or if one half of the Glass, or one half of the Pupil of the Eye be cover'd with any opake substance; the Object-glass, or that part of it or of the Eye which is not cover'd, may be consider'd as a Wedge with crooked Sides, and every Wedge of Glass or other pellucid Substance has the effect of a Prism in refracting the Light which passes through it.*

* *See our* Author's Lect. Optic. *Part* II. *Sect.* II. *pag. 269, &c.*

How the Colours in the ninth and tenth Experiments of the first Part arise from the different Reflexibility of Light, is evident by what was there said. But it is observable in the ninth Experiment, that whilst the Sun's direct Light is yellow, the Excess of the blue-making Rays in the reflected beam of Light MN, suffices only to bring that yellow to a pale white inclining to blue, and not to tinge it with

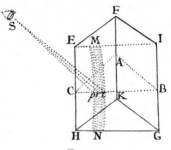

FIG. 13.

a manifestly blue Colour. To obtain therefore a better blue, I used instead of the yellow Light of the Sun the white Light of the Clouds, by varying a little the Experiment, as follows.

Exper. 16. Let HFG [in *Fig.* 13.] represent a Prism in the open Air, and S the Eye of the Spectator, viewing the Clouds by their Light coming into the Prism at the Plane Side FIGK, and reflected in it by its Base HEIG, and thence going out through its Plane Side HEFK to the Eye. And when the Prism and Eye are conveniently placed, so that the Angles of Incidence and Reflexion at the Base may be about

40 Degrees, the Spectator will see a Bow MN of a
blue Colour, running from one End of the Base to
the other, with the Concave Side towards him, and
the Part of the Base IMNG beyond this Bow will be
brighter than the other Part EMNH on the other
Side of it. This blue Colour MN being made by
nothing else than by Reflexion of a specular Superficies, seems so odd a Phænomenon, and so difficult
to be explained by the vulgar Hypothesis of Philosophers, that I could not but think it deserved to be
taken Notice of. Now for understanding the Reason
of it, suppose the Plane ABC to cut the Plane Sides
and Base of the Prism perpendicularly. From the Eye
to the Line BC, wherein that Plane cuts the Base,
draw the Lines Sp and St, in the Angles Spc 50 degr.
$\frac{1}{9}$, and Stc 49 degr. $\frac{1}{28}$, and the Point p will be the
Limit beyond which none of the most refrangible
Rays can pass through the Base of the Prism, and
be refracted, whose Incidence is such that they may
be reflected to the Eye; and the Point t will be the
like Limit for the least refrangible Rays, that is, beyond which none of them can pass through the Base,
whose Incidence is such that by Reflexion they may
come to the Eye. And the Point r taken in the middle
Way between p and t, will be the like Limit for the
meanly refrangible Rays. And therefore all the least
refrangible Rays which fall upon the Base beyond t,
that is, between t and B, and can come from thence
to the Eye, will be reflected thither: But on this side
t, that is, between t and c, many of these Rays will be
transmitted through the Base. And all the most re-

frangible Rays which fall upon the Base beyond p, that is, between p and B, and can by Reflexion come from thence to the Eye, will be reflected thither, but every where between p and c, many of these Rays will get through the Base, and be refracted; and the same is to be understood of the meanly refrangible Rays on either side of the Point r. Whence it follows, that the Base of the Prism must every where be-tween t and B, by a total Reflexion of all sorts of Rays to the Eye, look white and bright.   And every where between p and C, by reason of the Transmission of many Rays of every sort, look more pale, obscure, and dark. But at r, and in other Places between p and t, where all the more refrangible Rays are reflected to the Eye, and many of the less refrangible are trans-mitted, the Excess of the most refrangible in the re-flected Light will tinge that Light with their Colour, which is violet and blue. And this happens by taking the Line C prt B any where between the Ends of the Prism HG and EI.

PROP. IX.   PROB. IV.

By the discovered Properties of Light to explain the Colours of the Rain-bow.

THIS Bow never appears, but where it rains in the Sun-shine, and may be made artificially by spouting up Water which may break aloft, and scatter into Drops, and fall down like Rain. For the

Sun shining upon these Drops certainly causes the
Bow to appear to a Spectator standing in a due Position to the Rain and Sun. And hence it is now agreed
upon, that this Bow is made by Refraction of the
Sun's Light in drops of falling Rain. This was understood by some of the Antients, and of late more fully
discover'd and explain'd by the famous *Antonius de
Dominis* Archbishop of *Spalato*, in his book *De Radiis
Visûs & Lucis*, published by his Friend *Bartolus* at
Venice, in the Year 1611, and written above 20 Years
before. For he teaches there how the interior Bow is
made in round Drops of Rain by two Refractions of
the Sun's Light, and one Reflexion between them,
and the exterior by two Refractions, and two sorts
of Reflexions between them in each Drop of Water,
and proves his Explications by Experiments made
with a Phial full of Water, and with Globes of Glass
filled with Water, and placed in the Sun to make the
Colours of the two Bows appear in them. The same
Explication *Des-Cartes* hath pursued in his Meteors,
and mended that of the exterior Bow. But whilst they
understood not the true Origin of Colours, it's necessary to pursue it here a little farther. For understanding therefore how the Bow is made, let a Drop
of Rain, or any other spherical transparent Body be
represented by the Sphere BNFG, [in *Fig*. 14.] described with the Center C, and Semi-diameter CN.
And let AN be one of the Sun's Rays incident upon
it at N, and thence refracted to F, where let it either
go out of the Sphere by Refraction towards V, or be
reflected to G; and at G let it either go out by Re-

fraction to R, or be reflected to H; and at H let it go out by Refraction towards S, cutting the incident Ray in Y. Produce AN and RG, till they meet in X, and upon AX and NF, let fall the Perpendiculars CD and CE, and produce CD till it fall upon the

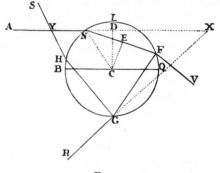

FIG. 14.

Circumference at L. Parallel to the incident Ray AN draw the Diameter BQ, and let the Sine of Incidence out of Air into Water be to the Sine of Refraction as I to R. Now, if you suppose the Point of Incidence N to move from the Point B, continually till it come to L, the Arch QF will first increase and then decrease, and so will the Angle AXR which the Rays AN and GR contain; and the Arch QF and Angle AXR will be biggest when ND is to CN as $\sqrt{II - RR}$ to $\sqrt{3RR}$, in which case NE will be to ND as 2R to I. Also the Angle AYS, which the Rays AN and HS contain will first decrease, and then increase and grow least when ND is to CN as $\sqrt{II - RR}$ to $\sqrt{8RR}$, in which case NE will be to ND, as 3R to I. And so the Angle which the next emergent Ray (that is, the emergent Ray after three Reflexions) contains with the incident Ray AN will come to its Limit when ND is to CN as $\sqrt{\overline{II - RR}}$ to $\sqrt{15}RR$, in which case NE will be to ND as 4R to I. And the Angle which the Ray next after that Emergent, that is, the Ray emergent after four Reflexions, contains with the Incident, will come to its Limit, when ND is to CN as $\sqrt{\overline{II - RR}}$ to $\sqrt{24}RR$, in which case NE will be to ND as 5R to I; and so on infinitely, the Numbers 3, 8, 15, 24, &c. being gather'd by continual Addition of the Terms of the arithmetical Progression 3, 5, 7, 9, &c. The Truth of all this Mathematicians will easily examine.*

Now it is to be observed, that as when the Sun comes to his Tropicks, Days increase and decrease but a very little for a great while together; so when by increasing the distance CD, these Angles come to their Limits, they vary their quantity but very little for some time together, and therefore a far greater number of the Rays which fall upon all the Points N in the Quadrant BL, shall emerge in the Limits of these Angles, than in any other Inclinations. And farther it is to be observed, that the Rays which differ in Refrangibility will have different Limits of their Angles of Emergence, and by consequence according to their different Degrees of Refrangibility emerge most copiously in different

* *This is demonstrated in our* Author's Lect. Optic. *Part.* I. *Sect.* IV. *Prop.* 35 *and* 36.

Angles, and being separated from one another appear each in their proper Colours. And what those Angles are may be easily gather'd from the foregoing Theorem by Computation.

For in the least refrangible Rays the Sines I and R (as was found above) are 108 and 81, and thence by Computation the greatest Angle AXR will be found 42 Degrees and 2 Minutes, and the least Angle AYS, 50 Degrees and 57 Minutes. And in the most refrangible Rays the Sines I and R are 109 and 81, and thence by Computation the greatest Angle AXR will be found 40 Degrees and 17 Minutes, and the least Angle AYS 54 Degrees and 7 Minutes.

Suppose now that O [in *Fig.* 15.] is the Spectator's Eye, and OP a Line drawn parallel to the Sun's Rays and let POE, POF, POG, POH, be Angles of 40 Degr. 17 Min. 42 Degr. 2 Min. 50 Degr. 57 Min. and 54 Degr. 7 Min. respectively, and these Angles turned about their common Side OP, shall with their other Sides OE, OF; OG, OH, describe the Verges of two Rain-bows AF, BE and CHDG. For if E, F, G, H, be drops placed any where in the conical Superficies described by OE, OF, OG, OH, and be illuminated by the Sun's Rays SE, SF, SG, SH; the Angle SEO being equal to the Angle POE, or 40 Degr. 17 Min. shall be the greatest Angle in which the most refrangible Rays can after one Reflexion be refracted to the Eye, and therefore all the Drops in the Line OE shall send the most refrangible Rays most copiously to the Eye, and thereby strike the Senses with the deepest violet Colour in that Region.

And in like manner the Angle SFO being equal to the Angle POF, or 42 Degr. 2 Min. shall be the greatest in which the least refrangible Rays after one Reflexion can emerge out of the Drops, and therefore those Rays shall come most copiously to the Eye

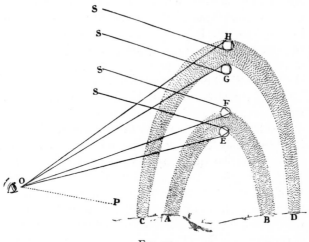

FIG. 15.

from the Drops in the Line OF, and strike the Senses with the deepest red Colour in that Region. And by the same Argument, the Rays which have intermediate Degrees of Refrangibility shall come most copiously from Drops between E and F, and strike the Senses with the intermediate Colours, in the Order which their Degrees of Refrangibility require, that is in the Progress from E to F, or from the inside of the Bow to the outside in this order, violet, indigo, blue, green, yellow, orange, red. But the violet, by the mixture of the white Light of the Clouds, will appear faint and incline to purple.

Again, the Angle SGO being equal to the Angle POG, or 50 Gr. 51 Min. shall be the least Angle in which the least refrangible Rays can after two Reflexions emerge out of the Drops, and therefore the least refrangible Rays shall come most copiously to the Eye from the Drops in the Line OG, and strike the Sense with the deepest red in that Region. And the Angle SHO being equal to the Angle POH, or 54 Gr. 7 Min. shall be the least Angle, in which the most refrangible Rays after two Reflexions can emerge out of the Drops; and therefore those Rays shall come most copiously to the Eye from the Drops in the Line OH, and strike the Senses with the deepest violet in that Region. And by the same Argument, the Drops in the Regions between G and H shall strike the Sense with the intermediate Colours in the Order which their Degrees of Refrangibility require, that is, in the Progress from G to H, or from the inside of the Bow to the outside in this order, red, orange, yellow, green, blue, indigo, violet. And since these four Lines OE, OF, OG, OH, may be situated any where in the above-mention'd conical Superficies; what is said of the Drops and Colours in these Lines is to be understood of the Drops and Colours every where in those Superficies.

Thus shall there be made two Bows of Colours, an interior and stronger, by one Reflexion in the Drops, and an exterior and fainter by two; for the Light be-

comes fainter by every Reflexion. And their Colours
shall lie in a contrary Order to one another, the red
of both Bows bordering upon the Space GF, which
is between the Bows. The Breadth of the interior
Bow EOF measured cross the Colours shall be 1
Degr. 45 Min. and the Breadth of the exterior GOH
shall be 3 Degr. 10 Min. and the distance between
them GOF shall be 8 Gr. 15 Min. the greatest Semi-
diameter of the innermost, that is, the Angle POF
being 42 Gr. 2 Min. and the least Semi-diameter of
the outermost POG, being 50 Gr. 57 Min. These are
the Measures of the Bows, as they would be were the
Sun but a Point; for by the Breadth of his Body, the
Breadth of the Bows will be increased, and their
Distance decreased by half a Degree, and so the
breadth of the interior Iris will be 2 Degr. 15 Min.
that of the exterior 3 Degr. 40 Min. their distance
8 Degr. 25 Min. the greatest Semi-diameter of the
interior Bow 42 Degr. 17 Min. and the least of the
exterior 50 Degr. 42 Min. And such are the Dimensions of the Bows in the Heavens found to be very
nearly, when their Colours appear strong and
perfect. For once, by such means as I then had, I
measured the greatest Semi-diameter of the interior
Iris about 42 Degrees, and the breadth of the red,
yellow and green in that Iris 63 or 64 Minutes, besides the outmost faint red obscured by the brightness of the Clouds, for which we may allow 3 or 4
Minutes more. The breadth of the blue was about
40 Minutes more besides the violet, which was so
much obscured by the brightness of the Clouds, that

I could not measure its breadth. But supposing the breadth of the blue and violet together to equal that of the red, yellow and green together, the whole breadth of this Iris will be about $2\frac{1}{4}$ Degrees, as above. The least distance between this Iris and the exterior Iris was about 8 Degrees and 30 Minutes. The exterior Iris was broader than the interior, but so faint, especially on the blue side, that I could not measure its breadth distinctly. At another time when both Bows appeared more distinct, I measured the breadth of the interior Iris 2 Gr. 10′, and the breadth of the red, yellow and green in the exterior Iris, was to the breadth of the same Colours in the interior as 3 to 2.

This Explication of the Rain-bow is yet farther confirmed by the known Experiment (made by *Antonius de Dominis* and *Des-Cartes*) of hanging up any where in the Sun-shine a Glass Globe filled with Water, and viewing it in such a posture, that the Rays which come from the Globe to the Eye may contain with the Sun's Rays an Angle of either 42 or 50 Degrees. For if the Angle be about 42 or 43 Degrees, the Spectator (suppose at O) shall see a full red Colour in that side of the Globe opposed to the Sun as 'tis represented at F, and if that Angle become less (suppose by depressing the Globe to E) there will appear other Colours, yellow, green and blue successive in the same side of the Globe. But if the Angle be made about 50 Degrees (suppose by lifting up the Globe to G) there will appear a red Colour in that side of the Globe towards the Sun, and if the Angle be made greater (suppose by lifting up the Globe to H) the red will turn successively to the other Colours, yellow, green and blue. The same thing I have tried, by letting a Globe rest, and raising or depressing the Eye, or otherwise moving it to make the Angle of a just magnitude.

I have heard it represented, that if the Light of a Candle be refracted by a Prism to the Eye; when the blue Colour falls upon the Eye, the Spectator shall see red in the Prism, and when the red falls upon the Eye he shall see blue; and if this were certain, the Colours of the Globe and Rain-bow ought to appear in a contrary order to what we find. But the Colours of the Candle being very faint, the mistake seems to arise from the difficulty of discerning what Colours fall on the Eye. For, on the contrary, I have sometimes had occasion to observe in the Sun's Light refracted by a Prism, that the Spectator always sees that Colour in the Prism which falls upon his Eye. And the same I have found true also in Candle-light. For when the Prism is moved slowly from the Line which is drawn directly from the Candle to the Eye, the red appears first in the Prism and then the blue, and therefore each of them is seen when it falls upon the Eye. For the red passes over the Eye first, and then the blue.

The Light which comes through drops of Rain by two Refractions without any Reflexion, ought to appear strongest at the distance of about 26 Degrees from the Sun, and to decay gradually both ways as the distance from him increases and decreases. And the same is to be understood of Light transmitted through spherical Hail-stones. And if the Hail be a little flatted, as it often is, the Light transmitted may grow so strong at a little less distance than that of 26 Degrees, as to form a Halo about the Sun or Moon; which Halo, as often as the Hail-stones are duly figured may be colour'd, and then it must be red within by the least refrangible Rays, and blue without by the most refrangible ones, especially if the Hail-stones have opake Globules of Snow in their center to intercept the Light within the Halo (as *Hugenius* has observ'd) and make the inside thereof more distinctly defined than it would otherwise be. For such Hail-stones, though spherical, by terminating the Light by the Snow, may make a Halo red within and colourless without, and darker in the red than without, as Halos used to be. For of those Rays which pass close by the Snow the Rubriform will be least refracted, and so come to the Eye in the directest Lines.

The Light which passes through a drop of Rain after two Refractions, and three or more Reflexions, is scarce strong enough to cause a sensible Bow; but in those Cylinders of Ice by which *Hugenius* explains the *Parhelia*, it may perhaps be sensible.

PROP. X. PROB. V.

By the discovered Properties of Light to explain the permanent Colours of Natural Bodies.

THESE Colours arise from hence, that some natural Bodies reflect some sorts of Rays, others other sorts more copiously than the rest. Minium reflects the least refrangible or red-making Rays most copiously, and thence appears red. Violets reflect the most refrangible most copiously, and thence have their Colour, and so of other Bodies. Every Body reflects the Rays of its own Colour more copiously than the rest, and from their excess and predominance in the reflected Light has its Colour.

Exper. 17. For if in the homogeneal Lights obtained by the solution of the Problem proposed in the fourth Proposition of the first Part of this Book, you place Bodies of several Colours, you will find, as I have done, that every Body looks most splendid and luminous in the Light of its own Colour. Cinnaber in the homogeneal red Light is most resplendent, in the green Light it is manifestly less resplendent, and in the blue Light still less. Indigo in the violet blue Light is most resplendent, and its splendor is gradually diminish'd, as it is removed thence by degrees through the green and yellow Light to the red. By a Leek the green Light, and next that the blue and yellow which compound green, are more strongly reflected than the other Colours red and violet, and so of the rest. But to make these Experiments the more manifest, such Bodies ought to be chosen as have the fullest and most vivid Colours, and two of those Bodies are to be compared together. Thus, for instance, if Cinnaber and *ultra*-marine blue, or some other full blue be held together in the red homogeneal Light, they will both appear red, but the Cinnaber will appear of a strongly luminous and resplendent red, and the *ultra*-marine blue of a faint obscure and dark red; and if they be held together in the blue homogeneal Light, they will both appear blue, but the *ultra*-marine will appear of a strongly luminous and resplendent blue, and the Cinnaber of a faint and dark blue. Which puts it out of dispute that the Cinnaber reflects the red Light much more copiously than the *ultra*-marine doth, and the *ultra*-marine reflects the blue Light much more copiously than the Cinnaber doth. The same Experiment may be tried successfully with red Lead and Indigo, or with any other two colour'd Bodies, if due allowance be made for the different strength or weakness of their Colour and Light.

And as the reason of the Colours of natural Bodies is evident by these Experiments, so it is farther confirmed and put past dispute by the two first Experiments of the first Part, whereby 'twas proved in such Bodies that the reflected Lights which differ in Colours do differ also in degrees of Refrangibility. For thence it's certain, that some Bodies reflect the more refrangible, others the less refrangible Rays more copiously.

And that this is not only a true reason of these

Colours, but even the only reason, may appear farther from this Consideration, that the Colour of homogeneal Light cannot be changed by the Reflexion of natural Bodies.

For if Bodies by Reflexion cannot in the least change the Colour of any one sort of Rays, they cannot appear colour'd by any other means than by reflecting those which either are of their own Colour, or which by mixture must produce it.

But in trying Experiments of this kind care must be had that the Light be sufficiently homogeneal. For if Bodies be illuminated by the ordinary prismatick Colours, they will appear neither of their own Day-light Colours, nor of the Colour of the Light cast on them, but of some middle Colour between both, as I have found by Experience. Thus red Lead (for instance) illuminated with the ordinary prismatick green will not appear either red or green, but orange or yellow, or between yellow and green, accordingly as the green Light by which 'tis illuminated is more or less compounded. For because red Lead appears red when illuminated with white Light, wherein all sorts of Rays are equally mix'd, and in the green Light all sorts of Rays are not equally mix'd, the Excess of the yellow-making, green-making and blue-making Rays in the incident green Light, will cause those Rays to abound so much in the reflected Light, as to draw the Colour from red towards their Colour. And because the red Lead reflects the red-making Rays most copiously in proportion to their number, and next after them the orange-making and yellow-making Rays; these Rays in the reflected Light will be more in proportion to the Light than they were in the incident green Light, and thereby will draw the reflected Light from green towards their Colour. And therefore the red Lead will appear neither red nor green, but of a Colour between both.

In transparently colour'd Liquors 'tis observable, that their Colour uses to vary with their thickness. Thus, for instance, a red Liquor in a conical Glass held between the Light and the Eye, looks of a pale and dilute yellow at the bottom where 'tis thin, and a little higher where 'tis thicker grows orange, and where 'tis still thicker becomes red, and where 'tis thickest the red is deepest and darkest. For it is to be conceiv'd that such a Liquor stops the indigo-making and violet-making Rays most easily, the blue-making Rays more difficultly, the green-making Rays still more difficultly, and the red-making most difficultly: And that if the thickness of the Liquor be only so much as suffices to stop a competent number of the violet-making and indigo-making Rays, without diminishing much the number of the rest, the rest must (by *Prop*. 6. *Part* 2.) compound a pale yellow. But if the Liquor be so much thicker as to stop also a great number of the blue-making Rays, and some of the green-making, the rest must compound an orange; and where it is so thick as to stop also a great number of the green-making and a considerable number of the yellow-making, the rest must begin to compound a red, and this red must grow deeper and darker as the yellow-making and orange-making Rays are more and more stopp'd by increasing the thickness of the Liquor, so that few Rays besides the red-making can get through.

Of this kind is an Experiment lately related to me by Mr. *Halley*, who, in diving deep into the Sea in a diving Vessel, found in a clear Sunshine Day, that when he was sunk many Fathoms deep into the Water the upper part of his Hand on which the Sun shone directly through the Water and through a small Glass Window in the Vessel appeared of a red Colour, like that of a Damask Rose, and the Water below and the under part of his Hand illuminated by Light reflected from the Water below look'd green. For thence it may be gather'd, that the Sea-Water reflects back the violet and blue-making Rays most easily, and lets the red-making Rays pass most freely and copiously to great Depths. For thereby the Sun's direct Light at all great Depths, by reason of the predominating red-making Rays, must appear red; and the greater the Depth is, the fuller and intenser must that red be. And at such Depths as the violet-making Rays scarce penetrate unto, the blue-making, green-making, and yellow-making Rays being reflected from below more copiously than the red-making ones, must compound a green.

Now, if there be two Liquors of full Colours, suppose a red and blue, and both of them so thick as suffices to make their Colours sufficiently full; though either Liquor be sufficiently transparent apart, yet will you not be able to see through both together. For, if only the red-making Rays pass through one Liquor, and only the blue-making through the other, no Rays can pass through both. This Mr. *Hook* tried casually with Glass Wedges filled with red and blue Liquors, and was surprized at the unexpected Event, the reason of it being then unknown; which makes me trust the more to his Experiment, though I have not tried it my self. But he that would repeat it, must take care the Liquors be of very good and full Colours.

Now, whilst Bodies become coloured by reflecting or transmitting this or that sort of Rays more copiously than the rest, it is to be conceived that they stop and stifle in themselves the Rays which they do not reflect or transmit. For, if Gold be foliated and held between your Eye and the Light, the Light looks of a greenish blue, and therefore massy Gold lets into its Body the blue-making Rays to be reflected to and fro within it till they be stopp'd and stifled, whilst it reflects the yellow-making outwards, and thereby looks yellow. And much after the same manner that Leaf Gold is yellow by reflected, and blue by transmitted Light, and massy Gold is yellow in all Positions of the Eye; there are some Liquors, as the Tincture of *Lignum Nephriticum*, and some sorts of Glass which transmit one sort of Light most copiously, and reflect another sort, and thereby look of several Colours, according to the Position of the Eye to the Light. But, if these Liquors or Glasses were so thick and massy that no Light could get through them, I question not but they would like all

other opake Bodies appear of one and the same
Colour in all Positions of the Eye, though this I
cannot yet affirm by Experience. For all colour'd
Bodies, so far as my Observation reaches, may be
seen through if made sufficiently thin, and therefore
are in some measure transparent, and differ only in
degrees of Transparency from tinged transparent
Liquors; these Liquors, as well as those Bodies, by a
sufficient Thickness becoming opake. A transparent
Body which looks of any Colour by transmitted
Light, may also look of the same Colour by reflected
Light, the Light of that Colour being reflected by
the farther Surface of the Body, or by the Air beyond
it. And then the reflected Colour will be diminished,
and perhaps cease, by making the Body very thick,
and pitching it on the backside to diminish the
Reflexion of its farther Surface, so that the Light
reflected from the tinging Particles may predominate.
In such Cases, the Colour of the reflected Light will
be apt to vary from that of the Light transmitted.
But whence it is that tinged Bodies and Liquors reflect some sort of Rays, and intromit or transmit
other sorts, shall be said in the next Book. In this
Proposition I content my self to have put it past
dispute, that Bodies have such Properties, and
thence appear colour'd.

PROP. XI.   Prob. VI.

*By mixing colour'd Lights to compound a beam of Light
of the same Colour and Nature with a beam of the
Sun's direct Light, and therein to experience the
Truth of the foregoing Propositions.*

LET ABC *abc* [in *Fig.* 16.] represent a Prism, by
which the Sun's Light let into a dark Chamber
through the Hole F, may be refracted towards the
Lens MN, and paint upon it at *p*, *q*, *r*, *s*, and *t*, the
usual Colours violet, blue, green, yellow, and red,
and let the diverging Rays by the Refraction of this
Lens converge again towards X, and there, by the
mixture of all those their Colours, compound a white
according to what was shewn above. Then let another Prism DEG *deg*, parallel to the former, be
placed at X, to refract that white Light upwards
towards Y. Let the refracting Angles of the Prisms,
and their distances from the Lens be equal, so that
the Rays which converged from the Lens towards X,
and without Refraction, would there have crossed
and diverged again, may by the Refraction of the
second Prism be reduced into Parallelism and diverge no more. For then those Rays will recompose
a beam of white Light XY. If the refracting Angle of
either Prism be the bigger, that Prism must be so
much the nearer to the Lens. You will know when
the Prisms and the Lens are well set together, by
observing if the beam of Light XY, which comes out
of the second Prism be perfectly white to the very

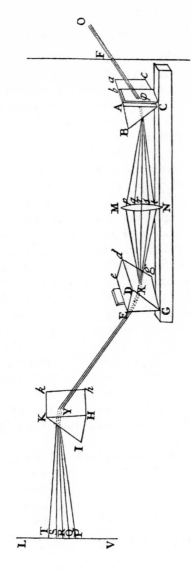

FIG. 16.

edges of the Light, and at all distances from the Prism continue perfectly and totally white like a beam of the Sun's Light. For till this happens, the Position of the Prisms and Lens to one another must be corrected; and then if by the help of a long beam of Wood, as is represented in the Figure, or by a Tube, or some other such Instrument, made for that Purpose, they be made fast in that Situation, you may try all the same Experiments in this compounded beam of Light XY, which have been made in the Sun's direct Light. For this compounded beam of Light has the same appearance, and is endow'd with all the same Properties with a direct beam of the Sun's Light, so far as my Observation reaches. And in trying Experiments in this beam you may by stopping any of the Colours, p, q, r, s, and t, at the Lens, see how the Colours produced in the Experiments are no other than those which the Rays had at the Lens before they entered the Composition of this Beam: And by consequence, that they arise not from any new Modifications of the Light by Refractions and Reflexions, but from the various Separations and Mixtures of the Rays originally endow'd with their colour-making Qualities.

So, for instance, having with a Lens $4\frac{1}{4}$ Inches broad, and two Prisms on either hand $6\frac{1}{4}$ Feet distant from the Lens, made such a beam of compounded Light; to examine the reason of the Colours made by Prisms, I refracted this compounded beam of Light XY with another Prism

HIK *kh*, and thereby cast the usual Prismatick
Colours PQRST upon the Paper LV placed behind.
And then by stopping any of the Colours *p, q, r, s, t*,
at the Lens, I found that the same Colour would
vanish at the Paper. So if the Purple *p* was stopp'd
at the Lens, the Purple P upon the Paper would
vanish, and the rest of the Colours would remain un-
alter'd, unless perhaps the blue, so far as some purple
latent in it at the Lens might be separated from it by
the following Refractions. And so by intercepting
the green upon the Lens, the green R upon the Paper
would vanish, and so of the rest; which plainly shews,
that as the white beam of Light XY was compounded
of several Lights variously colour'd at the Lens, so
the Colours which afterwards emerge out of it by
new Refractions are no other than those of which its
Whiteness was compounded. The Refraction of the
Prism HIK *kh* generates the Colours PQRST upon
the Paper, not by changing the colorific Qualities of
the Rays, but by separating the Rays which had the
very same colorific Qualities before they enter'd the
Composition of the refracted beam of white Light
XY. For otherwise the Rays which were of one
Colour at the Lens might be of another upon the
Paper, contrary to what we find.

So again, to examine the reason of the Colours of
natural Bodies, I placed such Bodies in the Beam
of Light XY, and found that they all appeared there
of those their own Colours which they have in Day-
light, and that those Colours depend upon the Rays
which had the same Colours at the Lens before they

enter'd the Composition of that beam. Thus, for
instance, Cinnaber illuminated by this beam appears
of the same red Colour as in Day-light ; and if at the
Lens you intercept the green-making and blue-
making Rays, its redness will become more full and
lively: But if you there intercept the red-making
Rays, it will not any longer appear red, but become
yellow or green, or of some other Colour, according
to the sorts of Rays which you do not intercept. So
Gold in this Light XY appears of the same yellow
Colour as in Day-light, but by intercepting at the
Lens a due Quantity of the yellow-making Rays it
will appear white like Silver (as I have tried) which
shews that its yellowness arises from the Excess of
the intercepted Rays tinging that Whiteness with
their Colour when they are let pass. So the Infusion
of *Lignum Nephriticum* (as I have also tried) when
held in this beam of Light XY, looks blue by the
reflected Part of the Light, and red by the transmitted Part of it, as when 'tis view'd in Day-light;
but if you intercept the blue at the Lens the Infusion
will lose its reflected blue Colour, whilst its transmitted red remains perfect, and by the loss of some
blue-making Rays, wherewith it was allay'd, becomes
more intense and full. And, on the contrary, if the
red and orange-making Rays be intercepted at the
Lens, the Infusion will lose its transmitted red,
whilst its blue will remain and become more full
and perfect. Which shews, that the Infusion does
not tinge the Rays with blue and red, but only
transmits those most copiously which were red-

making before, and reflects those most copiously which were blue-making before. And after the same manner may the Reasons of other Phænomena be examined, by trying them in this artificial beam of Light XY.

THE
SECOND BOOK
OF
OPTICKS

PART I.

Observations concerning the Reflexions, Refractions, and Colours of thin transparent Bodies.

IT has been observed by others, that transparent Substances, as Glass, Water, Air, &c. when made very thin by being blown into Bubbles, or otherwise formed into Plates, do exhibit various Colours according to their various thinness, altho' at a greater thickness they appear very clear and colourless. In the former Book I forbore to treat of these Colours, because they seemed of a more difficult Consideration, and were not necessary for establishing the Properties of Light there discoursed of. But because they may conduce to farther Discoveries for compleating the Theory of Light, especially as to the constitution of the parts of natural Bodies, on which their Colours or Transparency depend; I have here set down an account of them. To render this Discourse short and distinct, I have first described the principal of my Observations, and then consider'd and made use of them. The Observations are these.

Obs. 1. Compressing two Prisms hard together that their sides (which by chance were a very little convex) might somewhere touch one another: I found the place in which they touched to become absolutely transparent, as if they had there been one continued piece of Glass. For when the Light fell so obliquely on the Air, which in other places was between them, as to be all reflected; it seemed in that place of contact to be wholly transmitted, insomuch that when look'd upon, it appeared like a black or dark spot, by reason that little or no sensible Light was reflected from thence, as from other places; and when looked through it seemed (as it were) a hole in that Air which was formed into a thin Plate, by being compress'd between the Glasses. And through this hole Objects that were beyond might be seen distinctly, which could not at all be seen through other parts of the Glasses where the Air was interjacent. Although the Glasses were a little convex, yet this transparent spot was of a considerable breadth, which breadth seemed principally to proceed from the yielding inwards of the parts of the Glasses, by reason of their mutual pressure. For by pressing them very hard together it would become much broader than otherwise.

Obs. 2. When the Plate of Air, by turning the

Prisms about their common Axis, became so little
inclined to the incident Rays, that some of them
began to be transmitted, there arose in it many
slender Arcs of Colours which at first were shaped
almost like the Conchoid, as you see them delineated in the first Figure. And by continuing the
Motion of the Prisms, these Arcs increased and
bended more and more about the said transparent

Fig. 1.

spot, till they were compleated into Circles or Rings
incompassing it, and afterwards continually grew
more and more contracted.

These Arcs at their first appearance were of a
violet and blue Colour, and between them were
white Arcs of Circles, which presently by continuing
the Motion of the Prisms became a little tinged in
their inward Limbs with red and yellow, and to their
outward Limbs the blue was adjacent. So that the
order of these Colours from the central dark spot,
was at that time white, blue, violet; black, red,
orange, yellow, white, blue, violet, &c. But the
yellow and red were much fainter than the blue and
violet.

The Motion of the Prisms about their Axis being
continued, these Colours contracted more and more,

shrinking towards the whiteness on either side of it, until they totally vanished into it. And then the Circles in those parts appear'd black and white, without any other Colours intermix'd. But by farther moving the Prisms about, the Colours again emerged out of the whiteness, the violet and blue at its inward Limb, and at its outward Limb the red and yellow. So that now their order from the central Spot was white, yellow, red; black; violet, blue, white, yellow, red, &c. contrary to what it was before.

Obs. 3. When the Rings or some parts of them appeared only black and white, they were very distinct and well defined, and the blackness seemed as intense as that of the central Spot. Also in the Borders of the Rings, where the Colours began to emerge out of the whiteness, they were pretty distinct, which made them visible to a very great multitude. I have sometimes number'd above thirty Successions (reckoning every black and white Ring for one Succession) and seen more of them, which by reason of their smalness I could not number. But in other Positions of the Prisms, at which the Rings appeared of many Colours, I could not distinguish above eight or nine of them, and the Exterior of those were very confused and dilute.

In these two Observations to see the Rings distinct, and without any other Colour than Black and white, I found it necessary to hold my Eye at a good distance from them. For by approaching nearer, although in the same inclination of my Eye to the Plane of the Rings, there emerged a bluish Colour

out of the white, which by dilating it self more and
more into the black, render'd the Circles less distinct, and left the white a little tinged with red and
yellow. I found also by looking through a slit or
oblong hole, which was narrower than the pupil
of my Eye, and held close to it parallel to the Prisms,
I could see the Circles much distincter and visible
to a far greater number than otherwise.

Obs. 4. To observe more nicely the order of the
Colours which arose out of the white Circles as the
Rays became less and less inclined to the Plate of
Air; I took two Object-glasses, the one a Plano-
convex for a fourteen Foot Telescope, and the other
a large double Convex for one of about fifty Foot;
and upon this, laying the other with its plane side
downwards, I pressed them slowly together, to make
the Colours successively emerge in the middle of the
Circles, and then slowly lifted the upper Glass from
the lower to make them successively vanish again in
the same place. The Colour, which by pressing the
Glasses together, emerged last in the middle of the
other Colours, would upon its first appearance look
like a Circle of a Colour almost uniform from the
circumference to the center and by compressing the
Glasses still more, grow continually broader until a
new Colour emerged in its center, and thereby it became a Ring encompassing that new Colour. And by
compressing the Glasses still more, the diameter of
this Ring would increase, and the breadth of its
Orbit or Perimeter decrease until another new
Colour emerged in the center of the last: And so on

until a third, a fourth, a fifth, and other following new Colours successively emerged there, and became Rings encompassing the innermost Colour, the last of which was the black Spot. And, on the contrary, by lifting up the upper Glass from the lower, the diameter of the Rings would decrease, and the breadth of their Orbit increase, until their Colours reached successively to the center; and then they being of a considerable breadth, I could more easily discern and distinguish their Species than before. And by this means I observ'd their Succession and Quantity to be as followeth.

Next to the pellucid central Spot made by the contact of the Glasses succeeded blue, white, yellow, and red. The blue was so little in quantity, that I could not discern it in the Circles made by the Prisms, nor could I well distinguish any violet in it, but the yellow and red were pretty copious, and seemed about as much in extent as the white, and four or five times more than the blue. The next Circuit in order of Colours immediately encompassing these were violet, blue, green, yellow, and red: and these were all of them copious and vivid, excepting the green, which was very little in quantity, and seemed much more faint and dilute than the other Colours. Of the other four, the violet was the least in extent, and the blue less than the yellow or red. The third Circuit or Order was purple, blue, green, yellow, and red; in which the purple seemed more reddish than the violet in the former Circuit, and the green was much more conspicuous, being as brisk and copious as any of the other Colours, except the yellow, but the red began to be a little faded, inclining very much to purple. After this succeeded the fourth Circuit of green and red. The green was very copious and lively, inclining on the one side to blue, and on the other side to yellow. But in this fourth Circuit there was neither violet,

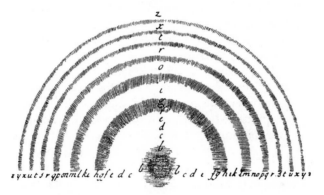

FIG. 2.

blue, nor yellow, and the red was very imperfect and dirty. Also the succeeding Colours became more and more imperfect and dilute, till after three or four revolutions they ended in perfect whiteness. Their form, when the Glasses were most compress'd so as to make the black Spot appear in the center, is delineated in the second Figure; where a, b, c, d, e:f, g, h, i, k: l, m, n, o, p: q, r: s, t: v, x: y, z, denote the Colours reckon'd in order from the center, black, blue, white, yellow, red: violet, blue, green, yellow, red: purple, blue, green, yellow, red: green, red: greenish blue, red: greenish blue, pale red: greenish blue, reddish white.

Obs. 5. To determine the interval of the Glasses, or thickness of the interjacent Air, by which each Colour was produced, I measured the Diameters of the first six Rings at the most lucid part of their Orbits, and squaring them, I found their Squares to be in the arithmetical Progression of the odd Numbers, 1, 3, 5, 7, 9, 11. And since one of these Glasses was plane, and the other spherical, their Intervals at those Rings must be in the same Progression. I measured also the Diameters of the dark or faint Rings between the more lucid Colours, and found their Squares to be in the arithmetical Progression of the even Numbers, 2, 4, 6, 8, 10, 12. And it being very nice and difficult to take these measures exactly; I repeated them divers times at divers parts of the Glasses, that by their Agreement I might be confirmed in them. And the same method I used in determining some others of the following Observations.

Obs. 6. The Diameter of the sixth Ring at the most lucid part of its Orbit was $\frac{58}{100}$ parts of an Inch, and the Diameter of the Sphere on which the double convex Object-glass was ground was about 102 Feet, and hence I gathered the thickness of the Air or Aereal Interval of the Glasses at that Ring. But some time after, suspecting that in making this Observation I had not determined the Diameter of the Sphere with sufficient accurateness, and being un-

certain whether the Plano-convex Glass was truly
plane, and not something concave or convex on that
side which I accounted plane; and whether I had not
pressed the Glasses together, as I often did, to make
them touch; (For by pressing such Glasses together
their parts easily yield inwards, and the Rings thereby
become sensibly broader than they would be, did the
Glasses keep their Figures.) I repeated the Experiment, and found the Diameter of the sixth lucid Ring
about $\frac{55}{100}$ parts of an Inch. I repeated the Experiment also with such an Object-glass of another Telescope as I had at hand. This was a double Convex
ground on both sides to one and the same Sphere,
and its Focus was distant from it $83\frac{2}{5}$ Inches. And
thence, if the Sines of Incidence and Refraction of
the bright yellow Light be assumed in proportion as
11 to 17, the Diameter of the Sphere to which the
Glass was figured will by computation be found 182
Inches. This Glass I laid upon a flat one, so that the
black Spot appeared in the middle of the Rings of
Colours without any other Pressure than that of the
weight of the Glass. And now measuring the Diameter of the fifth dark Circle as accurately as I could,
I found it the fifth part of an Inch precisely. This
Measure was taken with the points of a pair of Compasses on the upper Surface on the upper Glass, and
my Eye was about eight or nine Inches distance from
the Glass, almost perpendicularly over it, and the
Glass was $\frac{1}{6}$ of an Inch thick, and thence it is easy to
collect that the true Diameter of the Ring between
the Glasses was greater than its measur'd Diameter

above the Glasses in the Proportion of 80 to 79, or thereabouts, and by consequence equal to $\frac{16}{79}$ parts of an Inch, and its true Semi-diameter equal to $\frac{8}{79}$ parts. Now as the Diameter of the Sphere (182 Inches) is to the Semi-diameter of this fifth dark Ring ($\frac{8}{79}$ parts of an Inch) so is this Semi-diameter to the thickness of the Air at this fifth dark Ring; which is therefore $\frac{32}{567931}$ or $\frac{100}{1774784}$. Parts of an Inch; and the fifth Part thereof, *viz.* the $\frac{1}{88739}$ Part of an Inch, is the Thickness of the Air at the first of these dark Rings.

The same Experiment I repeated with another double convex Object-glass ground on both sides to one and the same Sphere. Its Focus was distant from it 168$\frac{1}{2}$ Inches, and therefore the Diameter of that Sphere was 184 Inches. This Glass being laid upon the same plain Glass, the Diameter of the fifth of the dark Rings, when the black Spot in their Center appear'd plainly without pressing the Glasses, was by the measure of the Compasses upon the upper Glass $\frac{121}{600}$ Parts of an Inch, and by consequence between the Glasses it was $\frac{1222}{6000}$: For the upper Glass was $\frac{1}{8}$ of an Inch thick, and my Eye was distant from it 8 Inches. And a third proportional to half this from the Diameter of the Sphere is $\frac{5}{88850}$ Parts of an Inch. This is therefore the Thickness of the Air at this Ring, and a fifth Part thereof, *viz.* the $\frac{1}{88850}$th Part of an Inch is the Thickness thereof at the first of the Rings, as above.

I tried the same Thing, by laying these Object-glasses upon flat Pieces of a broken Looking-glass, and found the same Measures of the Rings: Which makes me rely upon them till they can be determin'd more accurately by Glasses ground to larger Spheres, though in such Glasses greater care must be taken of a true Plane.

These Dimensions were taken, when my Eye was placed almost perpendicularly over the Glasses, being about an Inch, or an Inch and a quarter, distant from the incident Rays, and eight Inches distant from the Glass; so that the Rays were inclined to the Glass in an Angle of about four Degrees. Whence by the following Observation you will understand, that had the Rays been perpendicular to the Glasses, the Thickness of the Air at these Rings would have been less in the Proportion of the Radius to the Secant of four Degrees, that is, of 10000 to 10024. Let the Thicknesses found be therefore diminish'd in this Proportion, and they will become $\frac{1}{88952}$ and $\frac{1}{89063}$, or (to use the nearest round Number) the $\frac{1}{89000}$th Part of an Inch. This is the Thickness of the Air at the darkest Part of the first dark Ring made by perpendicular Rays; and half this Thickness multiplied by the Progression, 1, 3, 5, 7, 9, 11, &c. gives the Thicknesses of the Air at the most luminous Parts of all the brightest Rings, viz. $\frac{1}{178000}, \frac{3}{178000}, \frac{5}{178000}, \frac{7}{178000}$, &c. their arithmetical Means $\frac{2}{178000}, \frac{4}{178000}, \frac{6}{178000}$, &c. being its Thicknesses at the darkest Parts of all the dark ones.

Obs. 7. The Rings were least, when my Eye was placed perpendicularly over the Glasses in the Axis of the Rings: And when I view'd them obliquely they became bigger, continually swelling as I removed my Eye farther from the Axis. And partly by measuring the Diameter of the same Circle at several Obliquities of my Eye, partly by other Means, as also by making use of the two Prisms for very great Obliquities, I found its Diameter, and consequently the Thickness of the Air at its Perimeter in all those Obliquities to be very nearly in the Proportions express'd in this Table.

Angle of Incidence on the Air.		Angle of Refraction into the Air.		Diameter of the Ring.	Thickness of the Air.
Deg.	Min.				
00	00	00	00	10	10
06	26	10	00	$10\frac{1}{13}$	$10\frac{2}{13}$
12	45	20	00	$10\frac{1}{3}$	$10\frac{2}{3}$
18	49	30	00	$10\frac{3}{4}$	$11\frac{1}{2}$
24	30	40	00	$11\frac{2}{5}$	13
29	37	50	00	$12\frac{1}{2}$	$15\frac{1}{2}$
33	58	60	00	14	20
35	47	65	00	$15\frac{1}{4}$	$23\frac{1}{4}$
37	19	70	00	$16\frac{4}{5}$	$28\frac{1}{4}$
38	33	75	00	$19\frac{1}{4}$	37
39	27	80	00	$22\frac{6}{7}$	$52\frac{1}{4}$
40	00	85	00	29	$84\frac{1}{12}$
40	11	90	00	35	$122\frac{1}{2}$

In the two first Columns are express'd the Obliquities of the incident and emergent Rays to the Plate of the Air, that is, their Angles of Incidence

and Refraction. In the third Column the Diameter of
any colour'd Ring at those Obliquities is expressed
in Parts, of which ten constitute that Diameter when
the Rays are perpendicular. And in the fourth
Column the Thickness of the Air at the Circumference of that Ring is expressed in Parts, of which also
ten constitute its Thickness when the Rays are
perpendicular.

And from these Measures I seem to gather this
Rule: That the Thickness of the Air is proportional
to the Secant of an Angle, whose Sine is a certain
mean Proportional between the Sines of Incidence
and Refraction. And that mean Proportional, so far
as by these Measures I can determine it, is the first
of an hundred and six arithmetical mean Proportionals between those Sines counted from the bigger
Sine, that is, from the Sine of Refraction when the
Refraction is made out of the Glass into the Plate of
Air, or from the Sine of Incidence when the Refraction is made out of the Plate of Air into the Glass.

Obs. 8. The dark Spot in the middle of the Rings
increased also by the Obliquation of the Eye,
although almost insensibly. But, if instead of the
Object-glasses the Prisms were made use of, its Increase was more manifest when viewed so obliquely
that no Colours appear'd about it. It was least when
the Rays were incident most obliquely on the interjacent Air, and as the obliquity decreased it increased
more and more until the colour'd Rings appear'd, and
then decreased again, but not so much as it increased
before. And hence it is evident, that the Transpar-

ency was not only at the absolute Contact of the
Glasses, but also where they had some little Interval.
I have sometimes observed the Diameter of that Spot
to be between half and two fifth parts of the Diameter of the exterior Circumference of the red in
the first Circuit or Revolution of Colours when
view'd almost perpendicularly; whereas when view'd
obliquely it hath wholly vanish'd and become opake
and white like the other parts of the Glass; whence
it may be collected that the Glasses did then scarcely,
or not at all, touch one another, and that their Interval at the perimeter of that Spot when view'd
perpendicularly was about a fifth or sixth part of
their Interval at the circumference of the said red.

Observ. 9. By looking through the two contiguous
Object-glasses, I found that the interjacent Air ex-
hibited Rings of Colours, as well by transmitting
Light as by reflecting it. The central Spot was now
white, and from it the order of the Colours were
yellowish red; black, violet, blue, white, yellow, red;
violet, blue, green, yellow, red, *&c.* But these
Colours were very faint and dilute, unless when the
Light was trajected very obliquely through the
Glasses: For by that means they became pretty vivid.
Only the first yellowish red, like the blue in the
fourth Observation, was so little and faint as
scarcely to be discern'd. Comparing the colour'd
Rings made by Reflexion, with these made by transmission of the Light; I found that white was opposite to black, red to blue, yellow to violet, and
green to a Compound of red and violet. That is,

those parts of the Glass were black when looked through, which when looked upon appeared white, and on the contrary. And so those which in one case exhibited blue, did in the other case exhibit red. And the like of the other Colours. The manner you have represented in the third Figure, where AB, CD, are the Surfaces of the Glasses contiguous at E, and the black Lines between them are their Distances in arithmetical Progression, and the Colours written above are seen by reflected Light, and those below by Light transmitted (p. 209).

Obs. 10. Wetting the Object-glasses a little at their edges, the Water crept in slowly between them, and the Circles thereby became less and the Colours more faint: Insomuch that as the Water crept along, one half of them at which it first arrived would appear broken off from the other half, and contracted into a less Room. By measuring them I found the Proportions of their Diameters to the Diameters of the like Circles made by Air to be about seven to eight, and consequently the Intervals of the Glasses at like Circles, caused by those two Mediums Water and Air, are as about three to four. Perhaps it may be a general Rule, That if any other Medium more or less dense than Water be compress'd between the Glasses, their Intervals at the Rings caused thereby will be to their Intervals caused by interjacent Air, as the Sines are which measure the Refraction made out of that Medium into Air.

Obs. 11. When the Water was between the Glasses, if I pressed the upper Glass variously at its edges to make the Rings move nimbly from one place to another, a little white Spot would immediately follow the center of them, which upon creeping in of the ambient Water into that place would presently vanish. Its appearance was such as interjacent Air would have caused, and it exhibited the same Colours. But it was not air, for where any Bubbles of Air were in the Water they would not vanish. The Reflexion must have rather been caused by a subtiler Medium, which could recede through the Glasses at the creeping in of the Water.

Obs. 12. These Observations were made in the open Air. But farther to examine the Effects of colour'd Light falling on the Glasses, I darken'd the Room, and view'd them by Reflexion of the Colours of a Prism cast on a Sheet of white Paper, my Eye being so placed that I could see the colour'd Paper by Reflexion in the Glasses, as in a Looking-glass. And by this means the Rings became distincter and visible to a far greater number than in the open Air. I have sometimes seen more than twenty of them, whereas in the open Air I could not discern above eight or nine.

Obs. 13. Appointing an Assistant to move the Prism to and fro about its Axis, that all the Colours might successively fall on that part of the Paper which I saw by Reflexion from that part of the Glasses, where the Circles appear'd, so that all the Colours might be successively reflected from the Circles to my Eye, whilst I held it immovable, I found the Circles which the red Light made to be

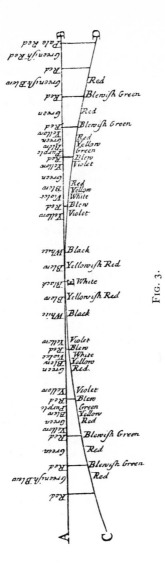

Fig. 3.

manifestly bigger than those which were made by the blue and violet. And it was very pleasant to see them gradually swell or contract accordingly as the Colour of the Light was changed. The Interval of the Glasses at any of the Rings when they were made by the utmost red Light, was to their Interval at the same Ring when made by the utmost violet, greater than as 3 to 2, and less than as 13 to 8. By the most of my Observations it was as 14 to 9. And this Proportion seem'd very nearly the same in all Obliquities of my Eye; unless when two Prisms were made use of instead of the Object-glasses. For then at a certain great obliquity of my Eye, the Rings made by the several Colours seem'd equal, and at a greater obliquity those made by the violet would be greater than the same Rings made by the red: the Refraction of the Prism in this case causing the most refrangible Rays to fall more obliquely on that plate of the Air than the least refrangible ones. Thus the Experiment succeeded in the colour'd Light, which was sufficiently strong and copious to make the Rings sensible. And thence it may be gather'd, that if the most refrangible and least refrangible Rays had been copious enough to make the Rings sensible without the mixture of other Rays, the Proportion which here was 14 to 9 would have been a little greater, suppose $14\frac{1}{4}$ or $14\frac{1}{3}$ to 9.

Obs. 14. Whilst the Prism was turn'd about its Axis with an uniform Motion, to make all the several Colours fall successively upon the Object-glasses, and thereby to make the Rings contract and

dilate: The Contraction or Dilatation of each Ring
thus made by the variation of its Colour was swiftest
in the red, and slowest in the violet, and in the intermediate Colours it had intermediate degrees of
Celerity. Comparing the quantity of Contraction and
Dilatation made by all the degrees of each Colour, I
found that it was greatest in the red; less in the
yellow, still less in the blue, and least in the violet.
And to make as just an Estimation as I could of the
Proportions of their Contractions or Dilatations, I
observ'd that the whole Contraction or Dilatation
of the Diameter of any Ring made by all the degrees
of red, was to that of the Diameter of the same Ring
made by all the degrees of violet, as about four to
three, or five to four, and that when the Light was
of the middle Colour between yellow and green, the
Diameter of the Ring was very nearly an arithmetical
Mean between the greatest Diameter of the same
Ring made by the outmost red, and the least Diameter thereof made by the outmost violet: Contrary
to what happens in the Colours of the oblong Spectrum made by the Refraction of a Prism, where the
red is most contracted, the violet most expanded,
and in the midst of all the Colours is the Confine of
green and blue. And hence I seem to collect that the
thicknesses of the Air between the Glasses there,
where the Ring is successively made by the limits of
the five principal Colours (red, yellow, green, blue,
violet) in order (that is, by the extreme red, by the
limit of red and yellow in the middle of the orange,
by the limit of yellow and green, by the limit of green

and blue, by the limit of blue and violet in the middle
of the indigo, and by the extreme violet) are to one
another very nearly as the sixth lengths of a Chord
which found the Notes in a sixth Major, *sol*, *la*, *mi*,
fa, *sol*, *la*. But it agrees something better with the
Observation to say, that the thicknesses of the Air
between the Glasses there, where the Rings are
successively made by the limits of the seven Colours,
red, orange, yellow, green, blue, indigo, violet in
order, are to one another as the Cube Roots of the
Squares of the eight lengths of a Chord, which found
the Notes in an eighth, *sol*, *la*, *fa*, *sol*, *la*, *mi*, *fa*, *sol*;
that is, as the Cube Roots of the Squares of the
Numbers, $1, \frac{8}{9}, \frac{5}{6}, \frac{3}{4}, \frac{2}{3}, \frac{3}{5}, \frac{9}{16}, \frac{1}{2}$.

Obs. 15. These Rings were not of various Colours
like those made in the open Air, but appeared all
over of that prismatick Colour only with which they
were illuminated. And by projecting the prismatick
Colours immediately upon the Glasses, I found that
the Light which fell on the dark Spaces which were
between the Colour'd Rings was transmitted through
the Glasses without any variation of Colour. For on
a white Paper placed behind, it would paint Rings
of the same Colour with those which were reflected,
and of the bigness of their immediate Spaces. And
from thence the origin of these Rings is manifest;
namely, that the Air between the Glasses, according to its various thickness, is disposed in some
places to reflect, and in others to transmit the Light
of any one Colour (as you may see represented in
the fourth Figure) and in the same place to reflect

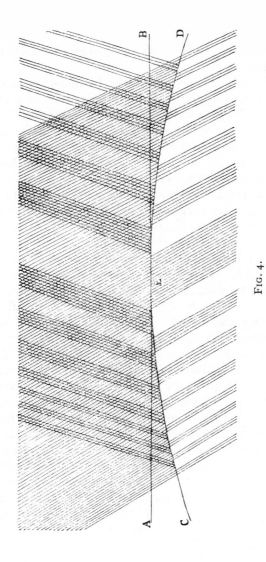

Fɪɢ. 4.

that of one Colour where it transmits that of another.

Obs. 16. The Squares of the Diameters of these Rings made by any prismatick Colour were in arithmetical Progression, as in the fifth Observation. And the Diameter of the sixth Circle, when made by the citrine yellow, and viewed almost perpendicularly was about $\frac{58}{100}$ parts of an Inch, or a little less, agreeable to the sixth Observation.

The precedent Observations were made with a rarer thin Medium, terminated by a denser, such as was Air or Water compress'd between two Glasses. In those that follow are set down the Appearances of a denser Medium thin'd within a rarer, such as are Plates of Muscovy Glass, Bubbles of Water, and some other thin Substances terminated on all sides with air.

Obs. 17. If a Bubble be blown with Water first made tenacious by dissolving a little Soap in it, 'tis a common Observation, that after a while it will appear tinged with a great variety of Colours. To defend these Bubbles from being agitated by the external Air (whereby their Colours are irregularly moved one among another, so that no accurate Observation can be made of them,) as soon as I had blown any of them I cover'd it with a clear Glass, and by that means its Colours emerged in a very regular order, like so many concentrick Rings encompassing the top of the Bubble. And as the Bubble grew thinner by the continual subsiding of the Water, these Rings dilated slowly and overspread the whole Bubble, descending in order to the bottom of it, where they vanish'd successively. In the mean while, after all the Colours were emerged at the top, there grew in the center of the Rings a small round black Spot, like that in the first Observation, which continually dilated it self till it became sometimes more than $\frac{1}{2}$ or $\frac{3}{4}$ of an Inch in breadth before the Bubble broke. At first I thought there had been no Light reflected from the Water in that place, but observing it more curiously, I saw within it several smaller round Spots, which appeared much blacker and darker than the rest, whereby I knew that there was some Reflexion at the other places which were not so dark as those Spots. And by farther Tryal I found that I could see the Images of some things (as of a Candle or the Sun) very faintly reflected, not only from the great black Spot, but also from the little darker Spots which were within it.

Besides the aforesaid colour'd Rings there would often appear small Spots of Colours, ascending and descending up and down the sides of the Bubble, by reason of some Inequalities in the subsiding of the Water. And sometimes small black Spots generated at the sides would ascend up to the larger black Spot at the top of the Bubble, and unite with it.

Obs. 18. Because the Colours of these Bubbles were more extended and lively than those of the Air thinn'd between two Glasses, and so more easy to be distinguish'd, I shall here give you a farther description of their order, as they were observ'd in viewing them by Reflexion of the Skies when of a white Colour, whilst a black substance was placed behind the Bubble. And they were these, red, blue; red, blue; red, blue; red, green; red, yellow, green, blue, purple; red, yellow, green, blue, violet; red, yellow, white, blue, black.

The three first Successions of red and blue were very dilute and dirty, especially the first, where the red seem'd in a manner to be white. Among these there was scarce any other Colour sensible besides red and blue, only the blues (and principally the second blue) inclined a little to green.

The fourth red was also dilute and dirty, but not so much as the former three; after that succeeded little or no yellow, but a copious green, which at first inclined a little to yellow, and then became a pretty brisk and good willow green, and afterwards changed to a bluish Colour; but there succeeded neither blue nor violet.

The fifth red at first inclined very much to purple, and afterwards became more bright and brisk, but yet not very pure. This was succeeded with a very bright and intense yellow, which was but little in quantity, and soon chang'd to green: But that green was copious and something more pure, deep and lively, than the former green. After that follow'd an excellent blue of a bright Sky-colour, and then a purple, which was less in quantity than the blue, and much inclined to red.

The sixth red was at first of a very fair and lively scarlet, and soon after of a brighter Colour, being very pure and brisk, and the best of all the reds. Then after a lively orange follow'd an intense bright

and copious yellow, which was also the best of all the
yellows, and this changed first to a greenish yellow,
and then to a greenish blue; but the green between
the yellow and the blue, was very little and dilute,
seeming rather a greenish white than a green. The
blue which succeeded became very good, and of a
very bright Sky-colour, but yet something inferior
to the former blue; and the violet was intense and
deep with little or no redness in it. And less in
quantity than the blue.

In the last red appeared a tincture of scarlet next
to violet, which soon changed to a brighter Colour,
inclining to an orange; and the yellow which follow'd
was at first pretty good and lively, but afterwards it
grew more dilute until by degrees it ended in perfect
whiteness. And this whiteness, if the Water was very
tenacious and well-temper'd, would slowly spread
and dilate it self over the greater part of the Bubble;
continually growing paler at the top, where at length
it would crack in many places, and those cracks, as
they dilated, would appear of a pretty good, but yet
obscure and dark Sky-colour; the white between the
blue Spots diminishing, until it resembled the Threds
of an irregular Net-work, and soon after vanish'd,
and left all the upper part of the Bubble of the said
dark blue Colour. And this Colour, after the aforesaid
manner, dilated it self downwards, until sometimes
it hath overspread the whole Bubble. In the mean
while at the top, which was of a darker blue than the
bottom, and appear'd also full of many round blue
Spots, something darker than the rest, there would

emerge one or more very black Spots, and within those, other Spots of an intenser blackness, which I mention'd in the former Observation; and these continually dilated themselves until the Bubble broke.

If the Water was not very tenacious, the black Spots would break forth in the white, without any sensible intervention of the blue. And sometimes they would break forth within the precedent yellow, or red, or perhaps within the blue of the second order, before the intermediate Colours had time to display themselves.

By this description you may perceive how great an affinity these Colours have with those of Air described in the fourth Observation, although set down in a contrary order, by reason that they begin to appear when the Bubble is thickest, and are most conveniently reckon'd from the lowest and thickest part of the Bubble upwards.

Obs. 19. Viewing in several oblique Positions of my Eye the Rings of Colours emerging on the top of the Bubble, I found that they were sensibly dilated by increasing the obliquity, but yet not so much by far as those made by thinn'd Air in the seventh Observation. For there they were dilated so much as, when view'd most obliquely, to arrive at a part of the Plate more than twelve times thicker than that where they appear'd when viewed perpendicularly; whereas in this case the thickness of the Water, at which they arrived when viewed most obliquely, was to that thickness which exhibited them by perpendicular Rays, something less than as 8 to 5. By the best of

my Observations it was between 15 and $15\frac{1}{2}$ to 10;
an increase about 24 times less than in the other case.

Sometimes the Bubble would become of an uniform thickness all over, except at the top of it near
the black Spot, as I knew, because it would exhibit
the same appearance of Colours in all Positions of the
Eye. And then the Colours which were seen at its
apparent circumference by the obliquest Rays,
would be different from those that were seen in other
places, by Rays less oblique to it. And divers Spec-
tators might see the same part of it of differing
Colours, by viewing it at very differing Obliquities.
Now observing how much the Colours at the same
places of the Bubble, or at divers places of equal
thickness, were varied by the several Obliquities of
the Rays; by the assistance of the 4th, 14th, 16th and
18th Observations, as they are hereafter explain'd,
I collect the thickness of the Water requisite to exhibit any one and the same Colour, at several Obliquities, to be very nearly in the Proportion expressed in this Table.

Incidence on the Water.		Refraction into the Water.		Thickness of the Water.
Deg.	*Min.*	*Deg.*	*Min.*	
00	00	00	00	10
15	00	11	11	$10\frac{1}{4}$
30	00	22	1	$10\frac{4}{5}$
45	00	32	2	$11\frac{4}{5}$
60	00	40	30	13
75	00	46	25	$14\frac{1}{2}$
90	00	48	35	$15\frac{1}{5}$

In the two first Columns are express'd the Obliquities of the Rays to the Superficies of the Water, that is, their Angles of Incidence and Refraction. Where I suppose, that the Sines which measure them are in round Numbers, as 3 to 4, though probably the Dissolution of Soap in the Water, may a little alter its refractive Virtue. In the third Column, the Thickness of the Bubble, at which any one Colour is exhibited in those several Obliquities, is express'd in Parts, of which ten constitute its Thickness when the Rays are perpendicular. And the Rule found by the seventh Observation agrees well with these Measures, if duly apply'd; namely, that the Thickness of a Plate of Water requisite to exhibit one and the same Colour at several Obliquities of the Eye, is proportional to the Secant of an Angle, whose Sine is the first of an hundred and six arithmetical mean Proportionals between the Sines of Incidence and Refraction counted from the lesser Sine, that is, from the Sine of Refraction when the Refraction is made out of Air into Water, otherwise from the Sine of Incidence.

I have sometimes observ'd, that the Colours which arise on polish'd Steel by heating it, or on Bell-metal, and some other metalline Substances, when melted and pour'd on the Ground, where they may cool in the open Air, have, like the Colours of Water-bubbles, been a little changed by viewing them at divers Obliquities, and particularly that a deep blue, or violet, when view'd very obliquely, hath been changed to a deep red. But the Changes of these

Colours are not so great and sensible as of those made by Water. For the Scoria, or vitrified Part of the Metal, which most Metals when heated or melted do continually protrude, and send out to their Surface, and which by covering the Metals in form of a thin glassy Skin, causes these Colours, is much denser than Water; and I find that the Change made by the Obliquation of the Eye is least in Colours of the densest thin Substances.

Obs. 20. As in the ninth Observation, so here, the Bubble, by transmitted Light, appear'd of a contrary Colour to that, which it exhibited by Reflexion. Thus when the Bubble being look'd on by the Light of the Clouds reflected from it, seemed red at its apparent Circumference, if the Clouds at the same time, or immediately after, were view'd through it, the Colour at its Circumference would be blue. And, on the contrary, when by reflected Light it appeared blue, it would appear red by transmitted Light.

Obs. 21. By wetting very thin Plates of *Muscovy* Glass, whose thinness made the like Colours appear, the Colours became more faint and languid, especially by wetting the Plates on that side opposite to the Eye: But I could not perceive any variation of their Species. So then the thickness of a Plate requisite to produce any Colour, depends only on the density of the Plate, and not on that of the ambient Medium. And hence, by the 10th and 16th Observations, may be known the thickness which Bubbles of Water, or Plates of *Muscovy* Glass, or other Substances, have at any Colour produced by them.

Obs. 22. A thin transparent Body, which is denser than its ambient Medium, exhibits more brisk and vivid Colours than that which is so much rarer; as I have particularly observed in the Air and Glass. For blowing Glass very thin at a Lamp Furnace, those Plates encompassed with Air did exhibit Colours much more vivid than those of Air made thin between two Glasses.

Obs. 23. Comparing the quantity of Light reflected from the several Rings, I found that it was most copious from the first or inmost, and in the exterior Rings became gradually less and less. Also the whiteness of the first Ring was stronger than that reflected from those parts of the thin Medium or Plate which were without the Rings; as I could manifestly perceive by viewing at a distance the Rings made by the two Object-glasses; or by comparing two Bubbles of Water blown at distant Times, in the first of which the Whiteness appear'd, which succeeded all the Colours, and in the other, the Whiteness which preceded them all.

Obs. 24. When the two Object-glasses were lay'd upon one another, so as to make the Rings of the Colours appear, though with my naked Eye I could not discern above eight or nine of those Rings, yet by viewing them through a Prism I have seen a far greater Multitude, insomuch that I could number more than forty, besides many others, that were so very small and close together, that I could not keep my Eye steady on them severally so as to number them, but by their Extent I have sometimes esti-

mated them to be more than an hundred. And I
believe the Experiment may be improved to the
Discovery of far greater Numbers. For they seem
to be really unlimited, though visible only so far as
they can be separated by the Refraction of the Prism,
as I shall hereafter explain.

But it was but one side of these Rings, namely,
that towards which the Refraction was made, which
by that Refraction was render'd distinct, and the
other side became more con-
fused than when view'd by
the naked Eye, insomuch that
there I could not discern above
one or two, and sometimes
none of those Rings, of which
I could discern eight or nine
with my naked Eye. And their
Segments or Arcs, which on
the other side appear'd so

FIG. 5.

numerous, for the most part exceeded not the third
Part of a Circle. If the Refraction was very great, or
the Prism very distant from the Object-glasses, the
middle Part of those Arcs became also confused, so
as to disappear and constitute an even Whiteness,
whilst on either side their Ends, as also the whole
Arcs farthest from the Center, became distincter
than before, appearing in the Form as you see them
design'd in the fifth Figure.

The Arcs, where they seem'd distinctest, were
only white and black successively, without any
other Colours intermix'd. But in other Places there

appeared Colours, whose Order was inverted by the refraction in such manner, that if I first held the Prism very near the Object-glasses, and then gradually removed it farther off towards my Eye, the Colours of the 2d, 3d, 4th, and following Rings, shrunk towards the white that emerged between them, until they wholly vanish'd into it at the middle of the Arcs, and afterwards emerged again in a contrary Order.   But at the Ends of the Arcs they retain'd their Order unchanged.

I have sometimes so lay'd one Object-glass upon the other, that to the naked Eye they have all over seem'd uniformly white, without the least Appearance of any of the colour'd Rings; and yet by viewing them through a Prism, great Multitudes of those Rings have discover'd themselves. And in like manner Plates of *Muscovy* Glass, and Bubbles of Glass blown at a Lamp-Furnace, which were not so thin as to exhibit any Colours to the naked Eye, have through the Prism exhibited a great Variety of them ranged irregularly up and down in the Form of Waves. And so Bubbles of Water, before they began to exhibit their Colours to the naked Eye of a By-stander, have appeared through a Prism, girded about with many parallel and horizontal Rings; to produce which Effect, it was necessary to hold the Prism parallel, or very nearly parallel to the Horizon, and to dispose it so that the Rays might be refracted upwards.

THE

SECOND BOOK

OF

OPTICKS

PART II.

Remarks upon the foregoing Observations.

HAVING given my Observations of these Colours, before I make use of them to unfold the Causes of the Colours of natural Bodies, it is convenient that by the simplest of them, such as are the 2d, 3d, 4th, 9th, 12th, 18th, 20th, and 24th, I first explain the more compounded. And first to shew how the Colours in the fourth and eighteenth Observations are produced, let there be taken in any Right Line from the Point Y, [in *Fig.* 6.] the Lengths YA, YB, YC, YD, YE, YF, YG, YH, in proportion to one another, as the Cube-Roots of the Squares of the Numbers, $\frac{1}{2}$, $\frac{9}{16}$, $\frac{3}{5}$, $\frac{2}{3}$, $\frac{3}{4}$, $\frac{5}{6}$, $\frac{8}{9}$, 1, whereby the Lengths of a Musical Chord to sound all the Notes in an eighth are represented; that is, in the Proportion of the Numbers 6300, 6814, 7114, 7631, 8255, 8855, 9243, 10000. And at the Points A, B, C, D, E, F, G, H, let Perpendiculars Aa, Bβ, &c. be erected, by whose Intervals the Extent of the several Colours set underneath against them, is to be represented. Then divide the Line Aa in such Proportion as the Numbers 1, 2, 3, 5, 6, 7, 9, 10, 11, &c. set at the Points of Division denote. And through those Divisions from Y draw Lines 1I, 2K, 3L, 5M, 6N, 7O, &c.

Now, if A2 be supposed to represent the Thickness of any thin transparent Body, at which the outmost Violet is most copiously reflected in the first Ring, or Series of Colours, then by the 13th Observation, HK will represent its Thickness, at which the utmost Red is most copiously reflected in the same Series. Also by the 5th and 16th Observations, A6 and HN will denote the Thicknesses at which those extreme Colours are most copiously reflected in the second Series, and A10 and HQ the Thicknesses at which they are most copiously reflected in the third Series, and so on. And the Thickness at which any of the intermediate Colours are reflected most copiously, will, according to the 14th Observation, be defined by the distance of the Line AH from the intermediate parts of the Lines 2K, 6N, 10Q, &c. against which the Names of those Colours are written below.

But farther, to define the Latitude of these Colours in each Ring or Series, let A1 design the least thickness, and A3 the greatest thickness, at which the

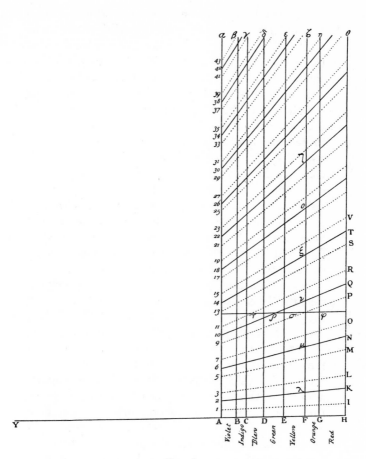

FIG. 6.

extreme violet in the first Series is reflected, and let HI, and HL, design the like limits for the extreme red, and let the intermediate Colours be limited by the intermediate parts of the Lines 1I, and 3L, against which the Names of those Colours are written, and so on: But yet with this caution, that the Reflexions be supposed strongest at the intermediate Spaces, 2K, 6N, 10Q, &c. and from thence to decrease gradually towards these limits, 1I, 3L, 5M, 7O, &c. on either side; where you must not conceive them to be precisely limited, but to decay indefinitely. And whereas I have assign'd the same Latitude to every Series, I did it, because although the Colours in the first Series seem to be a little broader than the rest, by reason of a stronger Reflexion there, yet that inequality is so insensible as scarcely to be determin'd by Observation.

Now according to this Description, conceiving that the Rays originally of several Colours are by turns reflected at the Spaces 1I, L3, 5M, O7, 9PR11, &c. and transmitted at the Spaces AHI1, 3LM5, 7OP9, &c. it is easy to know what Colour must in the open Air be exhibited at any thickness of a transparent thin Body. For if a Ruler be applied parallel to AH, at that distance from it by which the thickness of the Body is represented, the alternate Spaces 1IL3, 5MO7, &c. which it crosseth will denote the reflected original Colours, of which the Colour exhibited in the open Air is compounded. Thus if the constitution of the green in the third Series of Colours be desired, apply the Ruler as you

see at $\pi\varrho\sigma\varphi$, and by its passing through some of the
blue at π and yellow at σ, as well as through the green
at ϱ, you may conclude that the green exhibited at
that thickness of the Body is principally constituted
of original green, but not without a mixture of some
blue and yellow.

By this means you may know how the Colours
from the center of the Rings outward ought to
succeed in order as they were described in the 4th
and 18th Observations. For if you move the Ruler
gradually from AH through all distances, having
pass'd over the first Space which denotes little or no
Reflexion to be made by thinnest Substances, it will
first arrive at 1 the violet, and then very quickly at
the blue and green, which together with that violet
compound blue, and then at the yellow and red, by
whose farther addition that blue is converted into
whiteness, which whiteness continues during the
transit of the edge of the Ruler from I to 3, and after
that by the successive deficience of its component
Colours, turns first to compound yellow, and then to
red, and last of all the red ceaseth at L. Then begin
the Colours of the second Series, which succeed in
order during the transit of the edge of the Ruler from
5 to O, and are more lively than before, because
more expanded and severed. And for the same reason
instead of the former white there intercedes between
the blue and yellow a mixture of orange, yellow,
green, blue and indigo, all which together ought to
exhibit a dilute and imperfect green. So the Colours
of the third Series all succeed in order; first, the

violet, which a little interferes with the red of the second order, and is thereby inclined to a reddish purple ; then the blue and green, which are less mix'd with other Colours, and consequently more lively than before, especially the green: Then follows the yellow, some of which towards the green is distinct and good, but that part of it towards the succeeding red, as also that red is mix'd with the violet and blue of the fourth Series, whereby various degrees of red very much inclining to purple are compounded. This violet and blue, which should succeed this red, being mixed with, and hidden in it, there succeeds a green. And this at first is much inclined to blue, but soon becomes a good green, the only unmix'd and lively Colour in this fourth Series. For as it verges towards the yellow, it begins to interfere with the Colours of the fifth Series, by whose mixture the succeeding yellow and red are very much diluted and made dirty, especially the yellow, which being the weaker Colour is scarce able to shew it self. After this the several Series interfere more and more, and their Colours become more and more intermix'd, till after three or four more revolutions (in which the red and blue predominate by turns) all sorts of Colours are in all places pretty equally blended, and compound an even whiteness.

And since by the 15th Observation the Rays endued with one Colour are transmitted, where those of another Colour are reflected, the reason of the Colours made by the transmitted Light in the 9th and 20th Observations is from hence evident.

If not only the Order and Species of these Colours, but also the precise thickness of the Plate, or thin Body at which they are exhibited, be desired in parts of an Inch, that may be also obtained by assistance of the 6th or 16th Observations. For according to those Observations the thickness of the thinned Air, which between two Glasses exhibited the most luminous parts of the first six Rings were $\frac{1}{178000}, \frac{3}{178000},$ $\frac{5}{178000}, \frac{7}{178000}, \frac{9}{178000}, \frac{11}{178000}$ parts of an Inch. Suppose the Light reflected most copiously at these thicknesses be the bright citrine yellow, or confine of yellow and orange, and these thicknesses will be $F\lambda$, $F\mu$, Fv, $F\xi$, Fo, $F\tau$. And this being known, it is easy to determine what thickness of Air is represented by $G\varphi$, or by any other distance of the Ruler from AH.

But farther, since by the 10th Observation the thickness of Air was to the thickness of Water, which between the same Glasses exhibited the same Colour, as 4 to 3, and by the 21st Observation the Colours of thin Bodies are not varied by varying the ambient Medium; the thickness of a Bubble of Water, exhibiting any Colour, will be $\frac{3}{4}$ of the thickness of Air producing the same Colour. And so according to the same 10th and 21st Observations, the thickness of a Plate of Glass, whose Refraction of the mean refrangible Ray, is measured by the proportion of the Sines 31 to 20, may be $\frac{20}{31}$ of the thickness of Air producing the same Colours; and the like of other Mediums. I do not affirm, that this proportion of 20 to 31, holds in all the Rays; for the Sines of other sorts of Rays have other Proportions. But the differences of those Proportions are so little that I do not here consider them. On these Grounds I have composed the following Table, wherein the thickness of Air, Water, and Glass, at which each Colour is most intense and specifick, is expressed in parts of an Inch divided into ten hundred thousand equal parts.

Now if this Table be compared with the 6th Scheme, you will there see the constitution of each Colour, as to its Ingredients, or the original Colours of which it is compounded, and thence be enabled to judge of its Intenseness or Imperfection; which may suffice in explication of the 4th and 18th Observations, unless it be farther desired to delineate the manner how the Colours appear, when the two Object-glasses are laid upon one another. To do which, let there be described a large Arc of a Circle, and a streight Line which may touch that Arc, and parallel to that Tangent several occult Lines, at such distances from it, as the Numbers set against the several Colours in the Table denote. For the Arc, and its Tangent, will represent the Superficies of the Glasses terminating the interjacent Air; and the places where the occult Lines cut the Arc will show at what distances from the center, or Point of contact, each Colour is reflected.

There are also other Uses of this Table: For by its assistance the thickness of the Bubble in the 19th Observation was determin'd by the Colours which it exhibited. And so the bigness of the parts of natural Bodies may be conjectured by their Colours, as shall

The thickness of colour'd Plates and Particles of

		Air.	Water.	Glass.
Their Colours of the first Order,	Very black	$\frac{1}{2}$	$\frac{3}{8}$	$\frac{10}{31}$
	Black	1	$\frac{3}{4}$	$\frac{20}{31}$
	Beginning of Black	2	$1\frac{1}{2}$	$1\frac{2}{3}$
	Blue	$2\frac{2}{5}$	$1\frac{4}{5}$	$1\frac{11}{12}$
	White	$5\frac{1}{4}$	$3\frac{2}{5}$	$3\frac{2}{5}$
	Yellow	$7\frac{1}{9}$	$5\frac{3}{8}$	$4\frac{3}{5}$
	Orange	8	6	$5\frac{1}{6}$
	Red	9	$6\frac{3}{4}$	$5\frac{4}{5}$
Of the second order,	Violet	$11\frac{1}{6}$	$8\frac{3}{5}$	$7\frac{1}{5}$
	Indigo	$12\frac{5}{6}$	$9\frac{5}{8}$	$8\frac{2}{11}$
	Blue	14	$10\frac{1}{3}$	9
	Green	$15\frac{1}{8}$	$11\frac{2}{3}$	$9\frac{5}{7}$
	Yellow	$16\frac{2}{7}$	$12\frac{1}{2}$	$10\frac{2}{5}$
	Orange	$17\frac{2}{5}$	13	$11\frac{1}{6}$
	Bright red	$18\frac{1}{3}$	$13\frac{3}{4}$	$11\frac{5}{6}$
	Scarlet	$19\frac{2}{3}$	$14\frac{3}{4}$	$12\frac{2}{3}$
Of the third Order,	Purple	21	$15\frac{3}{4}$	$13\frac{11}{10}$
	Indigo	$22\frac{1}{10}$	$16\frac{4}{5}$	$14\frac{1}{4}$
	Blue	$23\frac{2}{5}$	$17\frac{11}{20}$	$15\frac{1}{10}$
	Green	$25\frac{1}{5}$	$18\frac{9}{10}$	$16\frac{1}{4}$
	Yellow	$27\frac{1}{7}$	$20\frac{2}{3}$	$17\frac{1}{2}$
	Red	29	$21\frac{3}{4}$	$18\frac{5}{8}$
	Bluish red	32	24	$20\frac{2}{3}$
Of the fourth Order,	Bluish green	34	$25\frac{1}{2}$	22
	Green	$35\frac{2}{7}$	$26\frac{1}{2}$	$22\frac{3}{4}$
	Yellowish green	36	27	$23\frac{4}{9}$
	Red	$40\frac{1}{3}$	$30\frac{1}{4}$	26
Of the fifth Order,	Greenish blue	46	$34\frac{1}{2}$	$29\frac{2}{3}$
	Red	$52\frac{1}{2}$	$39\frac{3}{8}$	34
Of the sixth Order,	Greenish blue	$58\frac{3}{4}$	44	38
	Red	65	$48\frac{3}{4}$	42
Of the seventh Order,	Greenish blue	71	$53\frac{1}{4}$	$45\frac{1}{5}$
	Ruddy White	77	$57\frac{3}{4}$	$49\frac{3}{5}$

be hereafter shewn. Also, if two or more very thin Plates be laid one upon another, so as to compose one Plate equalling them all in thickness, the resulting Colour may be hereby determin'd. For instance, Mr. *Hook* observed, as is mentioned in his *Micrographia*, that a faint yellow Plate of *Muscovy* Glass laid upon a blue one, constituted a very deep purple. The yellow of the first Order is a faint one, and the thickness of the Plate exhibiting it, according to the Table is $4\frac{3}{5}$, to which add 9, the thickness exhibiting blue of the second Order, and the Sum will be $13\frac{3}{5}$, which is the thickness exhibiting the purple of the third Order.

To explain, in the next place, the circumstances of the 2d and 3d Observations; that is, how the Rings of the Colours may (by turning the Prisms about their common Axis the contrary way to that expressed in those Observations) be converted into white and black Rings, and afterwards into Rings of Colours again, the Colours of each Ring lying now in an inverted order; it must be remember'd, that those Rings of Colours are dilated by the obliquation of the Rays to the Air which intercedes the Glasses, and that according to the Table in the 7th Observation, their Dilatation or Increase of their Diameter is most manifest and speedy when they are obliquest. Now the Rays of yellow being more refracted by the first Superficies of the said Air than those of red, are thereby made more oblique to the second Superficies, at which they are reflected to produce the colour'd Rings, and consequently the yellow Circle in each Ring will be more dilated than the red; and

the Excess of its Dilatation will be so much the
greater, by how much the greater is the obliquity of
the Rays, until at last it become of equal extent with
the red of the same Ring. And for the same reason
the green, blue and violet, will be also so much dilated by the still greater obliquity of their Rays, as
to become all very nearly of equal extent with the red,
that is, equally distant from the center of the Rings.
And then all the Colours of the same Ring must be
coincident, and by their mixture exhibit a white Ring.
And these white Rings must have black and dark
Rings between them, because they do not spread and
interfere with one another, as before. And for that
reason also they must become distincter, and visible
to far greater numbers. But yet the violet being
obliquest will be something more dilated, in proportion to its extent, than the other Colours, and so very
apt to appear at the exterior Verges of the white.

Afterwards, by a greater obliquity of the Rays, the
violet and blue become more sensibly dilated than
the red and yellow, and so being farther removed
from the center of the Rings, the Colours must
emerge out of the white in an order contrary to that
which they had before; the violet and blue at the exterior Limbs of each Ring, and the red and yellow at
the interior. And the violet, by reason of the greatest
obliquity of its Rays, being in proportion most of all
expanded, will soonest appear at the exterior Limb
of each white Ring, and become more conspicuous
than the rest. And the several Series of Colours belonging to the several Rings, will, by their unfolding

and spreading, begin again to interfere, and thereby render the Rings less distinct, and not visible to so great numbers.

If instead of the Prisms the Object-glasses be made use of, the Rings which they exhibit become not white and distinct by the obliquity of the Eye, by reason that the Rays in their passage through that Air which intercedes the Glasses are very nearly parallel to those Lines in which they were first incident on the Glasses, and consequently the Rays endued with several Colours are not inclined one more than another to that Air, as it happens in the Prisms.

There is yet another circumstance of these Experiments to be consider'd, and that is why the black and white Rings which when view'd at a distance appear distinct, should not only become confused by viewing them near at hand, but also yield a violet Colour at both the edges of every white Ring. And the reason is, that the Rays which enter the Eye at several parts of the Pupil, have several Obliquities to the Glasses, and those which are most oblique, if consider'd apart, would represent the Rings bigger than those which are the least oblique. Whence the breadth of the Perimeter of every white Ring is expanded outwards by the obliquest Rays, and inwards by the least oblique. And this Expansion is so much the greater by how much the greater is the difference of the Obliquity; that is, by how much the Pupil is wider, or the Eye nearer to the Glasses. And the breadth of the violet must be most expanded, because the Rays apt to

excite a Sensation of that Colour are most oblique
to a second or farther Superficies of the thinn'd
Air at which they are reflected, and have also the
greatest variation of Obliquity, which makes that
Colour soonest emerge out of the edges of the white.
And as the breadth of every Ring is thus augmented,
the dark Intervals must be diminish'd, until the
neighbouring Rings become continuous, and are
blended, the exterior first, and then those nearer the
center; so that they can no longer be distinguish'd
apart, but seem to constitute an even and uniform
whiteness.

Among all the Observations there is none accompanied with so odd circumstances as the twenty-
fourth. Of those the principal are, that in thin Plates,
which to the naked Eye seem of an even and uniform
transparent whiteness, without any terminations of
Shadows, the Refraction of a Prism should make
Rings of Colours appear, whereas it usually makes
Objects appear colour'd only there where they are
terminated with Shadows, or have parts unequally
luminous; and that it should make those Rings exceedingly distinct and white, although it usually
renders Objects confused and coloured. The Cause
of these things you will understand by considering,
that all the Rings of Colours are really in the Plate,
when view'd with the naked Eye, although by reason
of the great breadth of their Circumferences they so
much interfere and are blended together, that they
seem to constitute an uniform whiteness. But when
the Rays pass through the Prism to the Eye, the

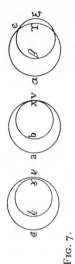

Fig. 7.

Orbits of the several Colours in every Ring are refracted, some more than others, according to their degrees of Refrangibility: By which means the Colours on one side of the Ring (that is in the circumference on one side of its center), become more unfolded and dilated, and those on the other side more complicated and contracted. And where by a due Refraction they are so much contracted, that the several Rings become narrower than to interfere with one another, they must appear distinct, and also white, if the constituent Colours be so much contracted as to be wholly coincident. But on the other side, where the Orbit of every Ring is made broader by the farther unfolding of its Colours, it must interfere more with other Rings than before, and so become less distinct.

To explain this a little farther, suppose the concentrick Circles AV, and BX, [in *Fig.* 7.] represent the red and violet of any Order, which, together with the intermediate Colours, constitute any one of these Rings. Now these being view'd through a Prism, the violet Circle BX, will, by a greater Refraction, be farther translated from its place than the red AV, and so approach nearer to it on that side of the Circles, towards which the Refractions are made. For instance, if the red be translated to av, the violet may be translated to bx, so as to approach nearer to it at x than before; and if the red be farther translated to av, the violet may be so much farther translated to bx as to convene with it at x; and if the red be yet farther translated to aY, the violet may be still so much farther translated to $\beta\xi$ as to pass beyond it at ξ, and convene with it at e and f. And this being understood not only of the red and violet, but of all the other intermediate Colours, and also of every revolution of those Colours, you will easily perceive how those of the same revolution or order, by their nearness at xv and $Y\xi$, and their coincidence at xv, e and f, ought to constitute pretty distinct Arcs of Circles, especially at xv, or at e and f; and that they will appear severally at xv, and at xv exhibit whiteness by their coincidence, and again appear severally at $Y\xi$, but yet in a contrary order to that which they had before, and still retain beyond e and f. But on the other side, at ab, ab, or $a\beta$, these Colours must become much more confused by being dilated and spread so as to interfere with those of other Orders. And the same confusion will happen at $Y\xi$ between e and f, if the Refraction be very great, or the Prism very distant from the Object-glasses: In which case no parts of the Rings will be seen, save only two little Arcs at e and f, whose distance from one another will be augmented by removing the Prism still farther from the Object-glasses: And these little Arcs must be distinctest and whitest at their middle, and at their ends, where they begin to grow confused, they must be colour'd. And the Colours at one end of every Arc must be in a contrary order to those at the other end, by reason that they cross in the intermediate white; namely, their ends, which verge towards $Y\xi$, will be red and yellow on that side next the center, and blue and violet on the other side. But their other ends which verge from $Y\xi$, will on the contrary be blue and violet on that side towards the center, and on the other side red and yellow.

Now as all these things follow from the properties of Light by a mathematical way of reasoning, so the truth of them may be manifested by Experiments. For in a dark Room, by viewing these Rings through a Prism, by reflexion of the several prismatick Colours, which an assistant causes to move to and fro upon a Wall or Paper from whence they are reflected, whilst the Spectator's Eye, the Prism, and the Object-glasses, (as in the 13th Observation,) are placed steady; the Position of the Circles made successively by the several Colours, will be found such, in respect of one another, as I have described in the Figures *abxv*, or abxv, or $\alpha\beta\xi Y$. And by the same method the truth of the Explications of other Observations may be examined.

By what hath been said, the like Phænomena of Water and thin Plates of Glass may be understood. But in small fragments of those Plates there is this farther observable, that where they lie flat upon a Table, and are turned about their centers whilst they

are view'd through a Prism, they will in some postures exhibit Waves of various Colours; and some of
them exhibit these Waves in one or two Positions
only, but the most of them do in all Positions exhibit
them, and make them for the most part appear almost
all over the Plates. The reason is, that the Superficies
of such Plates are not even, but have many Cavities
and Swellings, which, how shallow soever, do a little
vary the thickness of the Plate. For at the several
sides of those Cavities, for the Reasons newly described, there ought to be produced Waves in several
postures of the Prism. Now though it be but some
very small and narrower parts of the Glass, by which
these Waves for the most part are caused, yet they
may seem to extend themselves over the whole Glass,
because from the narrowest of those parts there are
Colours of several Orders, that is, of several Rings,
confusedly reflected, which by Refraction of the
Prism are unfolded, separated, and, according to
their degrees of Refraction, dispersed to several
places, so as to constitute so many several Waves, as
there were divers orders of Colours promiscuously
reflected from that part of the Glass.

These are the principal Phænomena of thin Plates
or Bubbles, whose Explications depend on the properties of Light, which I have heretofore deliver'd.
And these you see do necessarily follow from them,
and agree with them, even to their very least circumstances; and not only so, but do very much tend to
their proof. Thus, by the 24th Observation it appears,
that the Rays of several Colours, made as well by

thin Plates or Bubbles, as by Refractions of a Prism, have several degrees of Refrangibility; whereby those of each order, which at the reflexion from the Plate or Bubble are intermix'd with those of other orders, are separated from them by Refraction, and associated together so as to become visible by themselves like Arcs of Circles. For if the Rays were all alike refrangible, 'tis impossible that the whiteness, which to the naked Sense appears uniform, should by Refraction have its parts transposed and ranged into those black and white Arcs.

It appears also that the unequal Refractions of difform Rays proceed not from any contingent irregularities; such as are Veins, an uneven Polish, or fortuitous Position of the Pores of Glass; unequal and casual Motions in the Air or Æther, the spreading, breaking, or dividing the same Ray into many diverging parts; or the like. For, admitting any such irregularities, it would be impossible for Refractions to render those Rings so very distinct, and well defined, as they do in the 24th Observation. It is necessary therefore that every Ray have its proper and constant degree of Refrangibility connate with it, according to which its refraction is ever justly and regularly perform'd; and that several Rays have several of those degrees.

And what is said of their Refrangibility may be also understood of their Reflexibility, that is, of their Dispositions to be reflected, some at a greater, and others at a less thickness of thin Plates or Bubbles; namely, that those Dispositions are also connate with the Rays, and immutable; as may appear by the 13th, 14th, and 15th Observations, compared with the fourth and eighteenth.

By the Precedent Observations it appears also, that whiteness is a dissimilar mixture of all Colours, and that Light is a mixture of Rays endued with all those Colours. For, considering the multitude of the Rings of Colours in the 3d, 12th, and 24th Observations, it is manifest, that although in the 4th and 18th Observations there appear no more than eight or nine of those Rings, yet there are really a far greater number, which so much interfere and mingle with one another, as after those eight or nine revolutions to dilute one another wholly, and constitute an even and sensibly uniform whiteness. And consequently that whiteness must be allow'd a mixture of all Colours, and the Light which conveys it to the Eye must be a mixture of Rays endued with all those Colours.

But farther; by the 24th Observation it appears, that there is a constant relation between Colours and Refrangibility; the most refrangible Rays being violet, the least refrangible red, and those of inter-mediate Colours having proportionably intermediate degrees of Refrangibility. And by the 13th, 14th, and 15th Observations, compared with the 4th or 18th there appears to be the same constant relation be-tween Colour and Reflexibility; the violet being in like circumstances reflected at least thicknesses of any thin Plate or Bubble, the red at greatest thicknesses, and the intermediate Colours at intermediate thicknesses. Whence it follows, that the colorifick Dispositions of Rays are also connate with them, and immutable; and by consequence, that all the Productions and Appearances of Colours in the World are derived, not from any physical Change caused in Light by Refraction or Reflexion, but only from the various Mixtures or Separations of Rays, by virtue of their different Refrangibility or Reflexibility. And in this respect the Science of Colours becomes a Speculation as truly mathematical as any other part of Opticks. I mean, so far as they depend on the Nature of Light, and are not produced or alter'd by the Power of Imagination, or by striking or pressing the Eye.

THE

SECOND BOOK

OF

OPTICKS

PART III.

Of the permanent Colours of natural Bodies, and the
Analogy between them and the Colours of thin
transparent Plates.

I AM now come to another part of this Design,
which is to consider how the Phænomena of thin
transparent Plates stand related to those of all other
natural Bodies. Of these Bodies I have already told
you that they appear of divers Colours, accordingly
as they are disposed to reflect most copiously the
Rays originally endued with those Colours. But their
Constitutions, whereby they reflect some Rays more
copiously than others, remain to be discover'd; and
these I shall endeavour to manifest in the following
Propositions.

PROP. I.

Those Superficies of transparent Bodies reflect the greatest quantity of Light, which have the greatest refracting Power; that is, which intercede Mediums that differ most in their refractive Densities. And in the Confines of equally refracting Mediums there is no Reflexion.

THE Analogy between Reflexion and Refraction will appear by considering, that when Light passeth obliquely out of one Medium into another which refracts from the perpendicular, the greater is the difference of their refractive Density, the less Obliquity of Incidence is requisite to cause a total Reflexion. For as the Sines are which measure the Refraction, so is the Sine of Incidence at which the total Reflexion begins, to the Radius of the Circle; and consequently that Angle of Incidence is least where there is the greatest difference of the Sines. Thus in the passing of Light out of Water into Air, where the Refraction is measured by the Ratio of the Sines 3 to 4, the total Reflexion begins when the Angle of Incidence is about 48 Degrees 35 Minutes. In passing out of Glass into Air, where the Refraction is measured by the Ratio of the Sines 20 to 31, the total Reflexion begins when the Angle of Incidence is 40 Degrees 10 Minutes; and so in passing out of Crystal, or more strongly refracting Mediums into Air, there is still a less obliquity requisite to cause a total reflexion. Superficies therefore

which refract most do soonest reflect all the Light
which is incident on them, and so must be allowed
most strongly reflexive.

But the truth of this Proposition will farther
appear by observing, that in the Superficies interceding two transparent Mediums, (such as are Air,
Water, Oil, common Glass, Crystal, metalline
Glasses, Island Glasses, white transparent Arsenick,
Diamonds, &c.) the Reflexion is stronger or weaker
accordingly, as the Superficies hath a greater or less
refracting Power. For in the Confine of Air and Sal-
gem 'tis stronger than in the Confine of Air and
Water, and still stronger in the Confine of Air and
common Glass or Crystal, and stronger in the Confine of Air and a Diamond. If any of these, and such
like transparent Solids, be immerged in Water, its
Reflexion becomes much weaker than before; and
still weaker if they be immerged in the more strongly
refracting Liquors of well rectified Oil of Vitriol or
Spirit of Turpentine. If Water be distinguish'd into
two parts by any imaginary Surface, the Reflexion in
the Confine of those two parts is none at all. In the
Confine of Water and Ice 'tis very little; in that of
Water and Oil 'tis something greater; in that of
Water and Sal-gem still greater; and in that of Water
and Glass, or Crystal or other denser Substances
still greater, accordingly as those Mediums differ
more or less in their refracting Powers. Hence in the
Confine of common Glass and Crystal, there ought
to be a weak Reflexion, and a stronger Reflexion in
the Confine of common and metalline Glass; though

I have not yet tried this. But in the Confine of two Glasses of equal density, there is not any sensible Reflexion; as was shewn in the first Observation. And the same may be understood of the Superficies interceding two Crystals, or two Liquors, or any other Substances in which no Refraction is caused. So then the reason why uniform pellucid Mediums (such as Water, Glass, or Crystal,) have no sensible Reflexion but in their external Superficies, where they are adjacent to other Mediums of a different density, is because all their contiguous parts have one and the same degree of density.

Prop. II.

The least parts of almost all natural Bodies are in some measure transparent: And the Opacity of those Bodies ariseth from the multitude of Reflexions caused in their internal Parts.

THAT this is so has been observed by others, and will easily be granted by them that have been conversant with Microscopes. And it may be also tried by applying any substance to a hole through which some Light is immitted into a dark Room. For how opake soever that Substance may seem in the open Air, it will by that means appear very manifestly transparent, if it be of a sufficient thinness. Only white metalline Bodies must be excepted, which by reason of their excessive density seem to reflect almost all the Light incident on their first Superficies; unless by solution in Menstruums they be reduced into very small Particles, and then they become transparent.

PROP. III.

Between the parts of opake and colour'd Bodies are many Spaces, either empty, or replenish'd with Mediums of other Densities; as Water between the tinging Corpuscles wherewith any Liquor is impregnated, Air between the aqueous Globules that constitute Clouds or Mists; and for the most part Spaces void of both Air and Water, but yet perhaps not wholly void of all Substance, between the parts of hard Bodies.

THE truth of this is evinced by the two precedent Propositions: For by the second Proposition there are many Reflexions made by the internal parts of Bodies, which, by the first Proposition, would not happen if the parts of those Bodies were continued without any such Interstices between them; because Reflexions are caused only in Superficies, which intercede Mediums of a differing density, by *Prop.* 1.

But farther, that this discontinuity of parts is the principal Cause of the opacity of Bodies, will appear by considering, that opake Substances become transparent by filling their Pores with any Substance of equal or almost equal density with their parts. Thus Paper dipped in Water or Oil, the *Oculus Mundi* Stone steep'd in Water, Linnen Cloth oiled or varnish'd, and many other Substances soaked in such

Liquors as will intimately pervade their little Pores, become by that means more transparent than otherwise; so, on the contrary, the most transparent Substances, may, by evacuating their Pores, or separating their parts, be render'd sufficiently opake; as Salts or wet Paper, or the *Oculus Mundi* Stone by being dried, Horn by being scraped, Glass by being reduced to Powder, or otherwise flawed; Turpentine by being stirred about with Water till they mix imperfectly, and Water by being form'd into many small Bubbles, either alone in the form of Froth, or by shaking it together with Oil of Turpentine, or Oil Olive, or with some other convenient Liquor, with which it will not perfectly incorporate. And to the increase of the opacity of these Bodies, it conduces something, that by the 23d Observation the Reflexions of very thin transparent Substances are considerably stronger than those made by the same Substances of a greater thickness.

PROP. IV.

The Parts of Bodies and their Interstices must not be less than of some definite bigness, to render them opake and colour'd.

FOR the opakest Bodies, if their parts be subtilly divided, (as Metals, by being dissolved in acid Menstruums, &c.) become perfectly transparent. And you may also remember, that in the eighth Observation there was no sensible reflexion at the Superficies of the Object-glasses, where they were very

near one another, though they did not absolutely
touch. And in the 17th Observation the Reflexion of
the Water-bubble where it became thinnest was
almost insensible, so as to cause very black Spots to
appear on the top of the Bubble, by the want of reflected Light.

On these grounds I perceive it is that Water, Salt,
Glass, Stones, and such like Substances, are transparent. For, upon divers Considerations, they seem
to be as full of Pores or Interstices between their
parts as other Bodies are, but yet their Parts and
Interstices to be too small to cause Reflexions in their
common Surfaces.

PROP. V.

*The transparent parts of Bodies, according to their
several sizes, reflect Rays of one Colour, and transmit those of another, on the same grounds that thin
Plates or Bubbles do reflect or transmit those Rays.
And this I take to be the ground of all their Colours.*

FOR if a thinn'd or plated Body, which being of
an even thickness, appears all over of one uniform Colour, should be slit into Threads, or broken
into Fragments, of the same thickness with the Plate;
I see no reason why every Thread or Fragment
should not keep its Colour, and by consequence why
a heap of those Threads or Fragments should not
constitute a Mass or Powder of the same Colour,
which the Plate exhibited before it was broken. And
the parts of all natural Bodies being like so many

Fragments of a Plate, must on the same grounds exhibit the same Colours.

Now, that they do so will appear by the affinity of their Properties. The finely colour'd Feathers of some Birds, and particularly those of Peacocks Tails, do, in the very same part of the Feather, appear of several Colours in several Positions of the Eye, after the very same manner that thin Plates were found to do in the 7th and 19th Observations, and therefore their Colours arise from the thinness of the transparent parts of the Feathers; that is, from the slenderness of the very fine Hairs, or *Capillamenta*, which grow out of the sides of the grosser lateral Branches or Fibres of those Feathers. And to the same purpose it is, that the Webs of some Spiders, by being spun very fine, have appeared colour'd, as some have observ'd, and that the colour'd Fibres of some Silks, by varying the Position of the Eye, do vary their Colour. Also the Colours of Silks, Cloths, and other Substances, which Water or Oil can intimately penetrate, become more faint and obscure by being immerged in those Liquors, and recover their Vigor again by being dried; much after the manner declared of thin Bodies in the 10th and 21st Observations. Leaf-Gold, some sorts of painted Glass, the Infusion of *Lignum Nephriticum*, and some other Substances, reflect one Colour, and transmit another; like thin Bodies in the 9th and 20th Observations. And some of those colour'd Powders which Painters use, may have their Colours a little changed, by being very elaborately and finely ground. Where I see not

what can be justly pretended for those changes,
besides the breaking of their parts into less parts by
that contrition, after the same manner that the
Colour of a thin Plate is changed by varying its
thickness. For which reason also it is that the
colour'd Flowers of Plants and Vegetables, by being
bruised, usually become more transparent than before, or at least in some degree or other change their
Colours. Nor is it much less to my purpose, that, by
mixing divers Liquors, very odd and remarkable
Productions and Changes of Colours may be effected,
of which no cause can be more obvious and rational
than that the saline Corpuscles of one Liquor do
variously act upon or unite with the tinging Corpuscles of another, so as to make them swell, or
shrink, (whereby not only their bulk but their density also may be changed,) or to divide them into
smaller Corpuscles, (whereby a colour'd Liquor may
become transparent,) or to make many of them
associate into one cluster, whereby two transparent
Liquors may compose a colour'd one. For we see
how apt those saline Menstruums are to penetrate
and dissolve Substances to which they are applied,
and some of them to precipitate what others dissolve.
In like manner, if we consider the various Phænomena of the Atmosphere, we may observe, that when
Vapours are first raised, they hinder not the transparency of the Air, being divided into parts too
small to cause any Reflexion in their Superficies.
But when in order to compose drops of Rain they
begin to coalesce and constitute Globules of all inter-

mediate sizes, those Globules, when they become of
convenient size to reflect some Colours and transmit others, may constitute Clouds of various Colours
according to their sizes. And I see not what can be
rationally conceived in so transparent a Substance as
Water for the production of these Colours, besides
the various sizes of its fluid and globular Parcels.

Prop. VI.

*The parts of Bodies on which their Colours depend,
are denser than the Medium which pervades their
Interstices.*

THIS will appear by considering, that the Colour of a Body depends not only on the Rays
which are incident perpendicularly on its parts, but
on those also which are incident at all other Angles.
And that according to the 7th Observation, a very
little variation of obliquity will change the reflected
Colour, where the thin Body or small Particles is
rarer than the ambient Medium, insomuch that such
a small Particle will at diversly oblique Incidences
reflect all sorts of Colours, in so great a variety that
the Colour resulting from them all, confusedly reflected from a heap of such Particles, must rather be
a white or grey than any other Colour, or at best it
must be but a very imperfect and dirty Colour.
Whereas if the thin Body or small Particle be much
denser than the ambient Medium, the Colours, according to the 19th Observation, are so little changed
by the variation of obliquity, that the Rays which

are reflected least obliquely may predominate over the rest, so much as to cause a heap of such Particles to appear very intensely of their Colour.

It conduces also something to the confirmation of this Proposition, that, according to the 22d Observation, the Colours exhibited by the denser thin Body within the rarer, are more brisk than those exhibited by the rarer within the denser.

PROP. VII.

The bigness of the component parts of natural Bodies may be conjectured by their Colours.

FOR since the parts of these Bodies, by *Prop.* 5. do most probably exhibit the same Colours with a Plate of equal thickness, provided they have the same refractive density; and since their parts seem for the most part to have much the same density with Water or Glass, as by many circumstances is obvious to collect; to determine the sizes of those parts, you need only have recourse to the precedent Tables, in which the thickness of Water or Glass exhibiting any Colour is expressed. Thus if it be desired to know the diameter of a Corpuscle, which being of equal density with Glass shall reflect green of the third Order; the Number $16\frac{1}{4}$ shews it to be $\dfrac{16\frac{1}{4}}{10000}$ parts of an Inch.

The greatest difficulty is here to know of what Order the Colour of any Body is. And for this end we must have recourse to the 4th and 18th

Observations; from whence may be collected these particulars.

Scarlets, and other *reds*, *oranges*, and *yellows*, if they be pure and intense, are most probably of the second order. Those of the first and third order also may be pretty good; only the yellow of the first order is faint, and the orange and red of the third Order have a great Mixture of violet and blue.

There may be good *Greens* of the fourth Order, but the purest are of the third. And of this Order the green of all Vegetables seems to be, partly by reason of the Intenseness of their Colours, and partly because when they wither some of them turn to a greenish yellow, and others to a more perfect yellow or orange, or perhaps to red, passing first through all the aforesaid intermediate Colours. Which Changes seem to be effected by the exhaling of the Moisture which may leave the tinging Corpuscles more dense, and something augmented by the Accretion of the oily and earthy Part of that Moisture. Now the green, without doubt, is of the same Order with those Colours into which it changeth, because the Changes are gradual, and those Colours, though usually not very full, yet are often too full and lively to be of the fourth Order.

Blues and *Purples* may be either of the second or third Order, but the best are of the third. Thus the Colour of Violets seems to be of that Order, because their Syrup by acid Liquors turns red, and by urinous and alcalizate turns green. For since it is of the Nature of Acids to dissolve or attenuate, and of

Alcalies to precipitate or incrassate, if the Purple Colour of the Syrup was of the second Order, an acid Liquor by attenuating its tinging Corpuscles would change it to a red of the first Order, and an Alcali by incrassating them would change it to a green of the second Order; which red and green, especially the green, seem too imperfect to be the Colours produced by these Changes. But if the said Purple be supposed of the third Order, its Change to red of the second, and green of the third, may without any Inconvenience be allow'd.

If there be found any Body of a deeper and less reddish Purple than that of the Violets, its Colour most probably is of the second Order. But yet there being no Body commonly known whose Colour is constantly more deep than theirs, I have made use of their Name to denote the deepest and least reddish Purples, such as manifestly transcend their Colour in purity.

The *blue* of the first Order, though very faint and little, may possibly be the Colour of some Substances; and particularly the azure Colour of the Skies seems to be of this Order. For all Vapours when they begin to condense and coalesce into small Parcels, become first of that Bigness, whereby such an Azure must be reflected before they can constitute Clouds of other Colours. And so this being the first Colour which Vapours begin to reflect, it ought to be the Colour of the finest and most transparent Skies, in which Vapours are not arrived to that Grossness requisite to reflect other Colours, as we find it is by Experience.

Whiteness, if most intense and luminous, is that of the first Order, if less strong and luminous, a Mixture of the Colours of several Orders. Of this last kind is the Whiteness of Froth, Paper, Linnen, and most white Substances; of the former I reckon that of white Metals to be. For whilst the densest of Metals, Gold, if foliated, is transparent, and all Metals become transparent if dissolved in Menstruums or vitrified, the Opacity of white Metals ariseth not from their Density alone. They being less dense than Gold would be more transparent than it, did not some other Cause concur with their Density to make them opake. And this Cause I take to be such a Bigness of their Particles as fits them to reflect the white of the first order. For, if they be of other Thicknesses they may reflect other Colours, as is manifest by the Colours which appear upon hot Steel in tempering it, and sometimes upon the Surface of melted Metals in the Skin or Scoria which arises upon them in their cooling. And as the white of the first order is the strongest which can be made by Plates of transparent Substances, so it ought to be stronger in the denser Substances of Metals than in the rarer of Air, Water, and Glass. Nor do I see but that metallick Substances of such a Thickness as may fit them to reflect the white of the first order, may, by reason of their great Density (according to the Tenor of the first of these Propositions) reflect all the Light incident upon them, and so be as opake and splendent as it's possible for any Body to be. Gold, or Copper mix'd with less than half their Weight of

Silver, or Tin, or Regulus of Antimony, in fusion, or amalgamed with a very little Mercury, become white; which shews both that the Particles of white Metals have much more Superficies, and so are smaller, than those of Gold and Copper, and also that they are so opake as not to suffer the Particles of Gold or Copper to shine through them. Now it is scarce to be doubted but that the Colours of Gold and Copper are of the second and third order, and therefore the Particles of white Metals cannot be much bigger than is requisite to make them reflect the white of the first order. The Volatility of Mercury argues that they are not much bigger, nor may they be much less, lest they lose their Opacity, and become either transparent as they do when attenuated by Vitrification, or by Solution in Menstruums, or black as they do when ground smaller, by rubbing Silver, or Tin, or Lead, upon other Substances to draw black Lines. The first and only Colour which white Metals take by grinding their Particles smaller, is black, and therefore their white ought to be that which borders upon the black Spot in the Center of the Rings of Colours, that is, the white of the first order. But, if you would hence gather the Bigness of metallick Particles, you must allow for their Density. For were Mercury transparent, its Density is such that the Sine of Incidence upon it (by my Computation) would be to the Sine of its Refraction, as 71 to 20, or 7 to 2. And therefore the Thickness of its Particles, that they may exhibit the same Colours with those of Bubbles of Water, ought to be less than the Thickness of the

Skin of those Bubbles in the Proportion of 2 to 7. Whence it's possible, that the Particles of Mercury may be as little as the Particles of some transparent and volatile Fluids, and yet reflect the white of the first order.

Lastly, for the production of *black*, the Corpuscles must be less than any of those which exhibit Colours. For at all greater sizes there is too much Light reflected to constitute this Colour. But if they be supposed a little less than is requisite to reflect the white and very faint blue of the first order, they will, according to the 4th, 8th, 17th and 18th Observations, reflect so very little Light as to appear intensely black, and yet may perhaps variously refract it to and fro within themselves so long, until it happen to be stifled and lost, by which means they will appear black in all positions of the Eye without any transparency. And from hence may be understood why Fire, and the more subtile dissolver Putrefaction, by dividing the Particles of Substances, turn them to black, why small quantities of black Substances impart their Colour very freely and intensely to other Substances to which they are applied; the minute Particles of these, by reason of their very great number, easily overspreading the gross Particles of others; why Glass ground very elaborately with Sand on a Copper Plate, 'till it be well polish'd, makes the Sand, together with what is worn off from the Glass and Copper, become very black: why black Substances do soonest of all others become hot in the Sun's Light and burn, (which Effect may proceed partly from the multitude of Refractions in a little room, and partly from the easy Commotion of so very small Corpuscles;) and why blacks are usually a little inclined to a bluish Colour. For that they are so may be seen by illuminating white Paper by Light reflected from black Substances. For the Paper will usually appear of a bluish white; and the reason is, that black borders in the obscure blue of the order described in the 18th Observation, and therefore reflects more Rays of that Colour than of any other.

In these Descriptions I have been the more particular, because it is not impossible but that Miscroscopes may at length be improved to the discovery of the Particles of Bodies on which their Colours depend, if they are not already in some measure arrived to that degree of perfection. For if those Instruments are or can be so far improved as with sufficient distinctness to represent Objects five or six hundred times bigger than at a Foot distance they appear to our naked Eyes, I should hope that we might be able to discover some of the greatest of those Corpuscles. And by one that would magnify three or four thousand times perhaps they might all be discover'd, but those which produce blackness. In the mean while I see nothing material in this Discourse that may rationally be doubted of, excepting this Position: That transparent Corpuscles of the same thickness and density with a Plate, do exhibit the same Colour. And this I would have understood not without some Latitude, as well because those Corpuscles may be of irregular Figures, and many Rays must be obliquely incident on them, and so have a shorter way through them than the length of their Diameters, as because the straitness of the Medium put in on all sides within such Corpuscles may a little alter its Motions or other qualities on which the Reflexion depends. But yet I cannot much suspect the last, because I have observed of some small Plates of Muscovy Glass which were of an even thickness, that through a Microscope they have appeared of the same Colour at their edges and corners where the included Medium was terminated, which they appeared of in other places. However it will add much to our Satisfaction, if those Corpuscles can be discover'd with Microscopes; which if we shall at length attain to, I fear it will be the utmost improvement of this Sense. For it seems impossible to see the more secret and noble Works of Nature within the Corpuscles by reason of their transparency.

PROP. VIII.

The Cause of Reflexion is not the impinging of Light on the solid or impervious parts of Bodies, as is commonly believed.

THIS will appear by the following Considerations. First, That in the passage of Light out of Glass into Air there is a Reflexion as strong as in its passage out of Air into Glass, or rather a little stronger, and by many degrees stronger than in its passage out of Glass into Water. And it seems not probable that Air should have more strongly reflecting parts than Water or Glass. But if that should possibly be supposed, yet it will avail nothing; for the Reflexion is as strong or stronger when the Air is drawn away from the Glass, (suppose by the Air-Pump invented by *Otto Gueriet*, and improved and made useful by Mr. *Boyle*) as when it is adjacent to it. Secondly, If Light in its passage out of Glass into Air be incident more obliquely than at an Angle of 40 or 41 Degrees it is wholly reflected, if less obliquely it is in great measure transmitted. Now it is not to be imagined that Light at one degree of obliquity should meet with Pores enough in the Air to transmit the greater part of it, and at another degree of obliquity should meet with nothing but parts to reflect it wholly, especially considering that in its passage out of Air into Glass, how oblique soever be its Incidence, it finds Pores enough in the Glass to transmit a great part of it. If any Man suppose that it is not reflected by the Air, but by the outmost superficial parts of the Glass, there is still the same difficulty: Besides, that such a Supposition is unintelligible, and will also appear to be false by applying Water behind some part of the Glass instead of Air. For so in a convenient obliquity of the Rays, suppose of 45 or 46 Degrees, at which they are all reflected where the Air is adjacent to the Glass, they shall be in great measure transmitted where the Water is adjacent to it; which argues, that their Reflexion or Transmission depends on the constitution of the Air and Water behind the Glass, and not on the striking of the Rays upon the parts of the Glass.

Thirdly, If the Colours made by a Prism placed at the entrance of a Beam of Light into a darken'd Room be successively cast on a second Prism placed at a greater distance from the former, in such manner that they are all alike incident upon it, the second Prism may be so inclined to the incident Rays, that those which are of a blue Colour shall be all reflected by it, and yet those of a red Colour pretty copiously transmitted. Now if the Reflexion be caused by the parts of Air or Glass, I would ask, why at the same Obliquity of Incidence the blue should wholly impinge on those parts, so as to be all reflected, and yet the red find Pores enough to be in a great measure transmitted. Fourthly, Where two Glasses touch one another, there is no sensible Reflexion, as was declared in the first Observation; and yet I see no reason why the Rays should not impinge on the parts of Glass, as much when contiguous to other Glass as when contiguous to Air. Fifthly, When the top of a Water-Bubble (in the 17th Observation,) by the continual subsiding and exhaling of the Water grew very thin, there was such a little and almost insensible quantity of Light reflected from it, that it appeared intensely black; whereas round about that black Spot, where the Water was thicker, the Reflexion was so strong as to make the Water seem very white. Nor is it only at the least thickness of thin Plates or Bubbles, that there is no manifest Reflexion, but at many other thicknesses continually greater and greater. For in the 15th Observation the Rays of the same Colour were by turns transmitted at one thick-

ness, and reflected at another thickness, for an inde-
terminate number of Successions. And yet in the
Superficies of the thinned Body, where it is of any
one thickness, there are as many parts for the Rays
to impinge on, as where it is of any other thickness.
Sixthly, If Reflexion were caused by the parts of reflecting Bodies, it would be impossible for thin Plates
or Bubbles, at one and the same place, to reflect the
Rays of one Colour, and transmit those of another,
as they do according to the 13th and 15th Observations. For it is not to be imagined that at one place
the Rays which, for instance, exhibit a blue Colour,
should have the fortune to dash upon the parts, and
those which exhibit a red to hit upon the Pores of
the Body; and then at another place, where the Body
is either a little thicker or a little thinner, that on the
contrary the blue should hit upon its pores, and
the red upon its parts. Lastly, Were the Rays of
Light reflected by impinging on the solid parts of
Bodies, their Reflexions from polish'd Bodies could
not be so regular as they are. For in polishing Glass
with Sand, Putty, or Tripoly, it is not to be imagined
that those Substances can, by grating and fretting
the Glass, bring all its least Particles to an accurate
Polish; so that all their Surfaces shall be truly plain
or truly spherical, and look all the same way, so as
together to compose one even Surface. The smaller
the Particles of those Substances are, the smaller will
be the Scratches by which they continually fret and
wear away the Glass until it be polish'd; but be they
never so small they can wear away the Glass no

otherwise than by grating and scratching it, and breaking the Protuberances; and therefore polish it no otherwise than by bringing its roughness to a very fine Grain, so that the Scratches and Frettings of the Surface become too small to be visible. And therefore if Light were reflected by impinging upon the solid parts of the Glass, it would be scatter'd as much by the most polish'd Glass as by the roughest. So then it remains a Problem, how Glass polish'd by fretting Substances can reflect Light so regularly as it does. And this Problem is scarce otherwise to be solved, than by saying, that the Reflexion of a Ray is effected, not by a single point of the reflecting Body, but by some power of the Body which is evenly diffused all over its Surface, and by which it acts upon the Ray without immediate Contact. For that the parts of Bodies do act upon Light at a distance shall be shewn hereafter.

Now if Light be reflected, not by impinging on the solid parts of Bodies, but by some other principle; it's probable that as many of its Rays as impinge on the solid parts of Bodies are not reflected but stifled and lost in the Bodies. For otherwise we must allow two sorts of Reflexions. Should all the Rays be reflected which impinge on the internal parts of clear Water or Crystal, those Substances would rather have a cloudy Colour than a clear Transparency. To make Bodies look black, it's necessary that many Rays be stopp'd, retained, and lost in them; and it seems not probable that any Rays can be

stopp'd and stifled in them which do not impinge on
their parts.

And hence we may understand that Bodies are
much more rare and porous than is commonly believed. Water is nineteen times lighter, and by consequence nineteen times rarer than Gold; and Gold is
so rare as very readily and without the least opposition to transmit the magnetick Effluvia, and easily to
admit Quicksilver into its Pores, and to let Water
pass through it. For a concave Sphere of Gold filled
with Water, and solder'd up, has, upon pressing the
Sphere with great force, let the Water squeeze
through it, and stand all over its outside in multitudes of small Drops, like Dew, without bursting or
cracking the Body of the Gold, as I have been inform'd by an Eye witness. From all which we may
conclude, that Gold has more Pores than solid parts,
and by consequence that Water has above forty times
more Pores than Parts. And he that shall find out an
Hypothesis, by which Water may be so rare, and yet
not be capable of compression by force, may doubt-
less by the same Hypothesis make Gold, and Water,
and all other Bodies, as much rarer as he pleases; so
that Light may find a ready passage through transparent Substances.

The Magnet acts upon Iron through all dense
Bodies not magnetick nor red hot, without any diminution of its Virtue; as for instance, through Gold,
Silver, Lead, Glass, Water. The gravitating Power
of the Sun is transmitted through the vast Bodies of
the Planets without any diminution, so as to act upon

all their parts to their very centers with the same
Force and according to the same Laws, as if the part
upon which it acts were not surrounded with the
Body of the Planet, The Rays of Light, whether they
be very small Bodies projected, or only Motion or
Force propagated, are moved in right Lines; and
whenever a Ray of Light is by any Obstacle turned
out of its rectilinear way, it will never return into the
same rectilinear way, unless perhaps by very great
accident. And yet Light is transmitted through pellucid solid Bodies in right Lines to very great distances.
How Bodies can have a sufficient quantity of Pores
for producing these Effects is very difficult to conceive, but perhaps not altogether impossible. For
the Colours of Bodies arise from the Magnitudes of
the Particles which reflect them, as was explained
above. Now if we conceive these Particles of Bodies
to be so disposed amongst themselves, that the Intervals or empty Spaces between them may be equal
in magnitude to them all; and that these Particles
may be composed of other Particles much smaller,
which have as much empty Space between them as
equals all the Magnitudes of these smaller Particles:
And that in like manner these smaller Particles are
again composed of others much smaller, all which
together are equal to all the Pores or empty Spaces
between them; and so on perpetually till you come
to solid Particles, such as have no Pores or empty
Spaces within them: And if in any gross Body there
be, for instance, three such degrees of Particles, the
least of which are solid; this Body will have seven

times more Pores than solid Parts. But if there be
four such degrees of Particles, the least of which are
solid, the Body will have fifteen times more Pores
than solid Parts. If there be five degrees, the Body
will have one and thirty times more Pores than solid
Parts. If six degrees, the Body will have sixty and
three times more Pores than solid Parts. And so on
perpetually. And there are other ways of conceiving
how Bodies may be exceeding porous. But what is
really their inward Frame is not yet known to us.

PROP. IX.

Bodies reflect and refract Light by one and the same
power, variously exercised in various Circum-
stances.

THIS appears by several Considerations. First,
Because when Light goes out of Glass into
Air, as obliquely as it can possibly do. If its Incidence
be made still more oblique, it becomes totally reflected. For the power of the Glass after it has refracted the Light as obliquely as is possible, if the
Incidence be still made more oblique, becomes too
strong to let any of its Rays go through, and by consequence causes total Reflexions. Secondly, Because
Light is alternately reflected and transmitted by thin
Plates of Glass for many Successions, accordingly
as the thickness of the Plate increases in an arithmetical Progression. For here the thickness of the
Glass determines whether that Power by which Glass
acts upon Light shall cause it to be reflected, or

suffer it to be transmitted. And, Thirdly, because those Surfaces of transparent Bodies which have the greatest refracting power, reflect the greatest quantity of Light, as was shewn in the first Proposition.

PROP. X.

If Light be swifter in Bodies than in Vacuo, in the proportion of the Sines which measure the Refraction of the Bodies, the Forces of the Bodies to reflect and refract Light, are very nearly proportional to the densities of the same Bodies; excepting that unctuous and sulphureous Bodies refract more than others of this same density.

LET AB represent the refracting plane Surface of any Body, and IC a Ray incident very obliquely upon the Body in C, so that the Angle ACI may be infinitely little, and let CR be the refracted Ray.

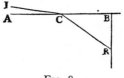

FIG. 8.

From a given Point B perpendicular to the refracting Surface erect BR meeting with the refracting Ray CR in R, and if CR represent the Motion of the refracted Ray, and this Motion be distinguish'd into two Motions CB and BR, whereof CB is parallel to the refracting Plane, and BR perpendicular to it: CB shall represent the Motion of the incident Ray, and BR the Motion generated by the Refraction, as Opticians have of late explain'd.

Now if any Body or Thing, in moving through any Space of a given breadth terminated on both sides by two parallel Planes, be urged forward in all parts of that Space by Forces tending directly forwards towards the last Plane, and before its Incidence on the first Plane, had no Motion towards it, or but an infinitely little one; and if the Forces in all parts of that Space, between the Planes, be at equal distances from the Planes equal to one another, but at several distances be bigger or less in any given Proportion, the Motion generated by the Forces in the whole passage of the Body or thing through that Space shall be in a subduplicate Proportion of the Forces, as Mathematicians will easily understand. And therefore, if the Space of activity of the refracting Superficies of the Body be consider'd as such a Space, the Motion of the Ray generated by the refracting Force of the Body, during its passage through that Space, that is, the Motion BR, must be in subduplicate Proportion of that refracting Force. I say therefore, that the Square of the Line BR, and by consequence the refracting Force of the Body, is very nearly as the density of the same Body. For this will appear by the following Table, wherein the Proportion of the Sines which measure the Refractions of several Bodies, the Square of BR, supposing CB an unite, the Densities of the Bodies estimated by their Specifick Gravities, and their Refractive Power in respect of their Densities are set down in several Columns.

The refracting Bodies.	The Proportion of the Sines of Incidence and Refraction of yellow Light.		The Square of BR, to which the refracting force of the Body is proportionate.	The density and specifick gravity of the Body.	The refractive Power of the Body in respect of its density.
A Pseudo-Topazius, being a natural, pellucid, brittle, hairy Stone, of a yellow Colour.	23 to	14	1'699	4'27	3979
Air.	3201 to	3200	0'000625	0'0012	5208
Glass of Antimony.	17 to	9	2'568	5'28	4864
A Selenitis.	61 to	41	1'213	2'252	5386
Glass vulgar.	31 to	20	1'4025	2'58	5436
Crystal of the Rock.	25 to	16	1'445	2'65	5450
Island Crystal.	5 to	3	1'778	2'72	6536
Sal Gemmæ.	17 to	11	1'388	2'143	6477
Alume.	35 to	24	1'1267	1'714	6570
Borax.	22 to	15	1'1511	1'714	6716
Niter.	32 to	21	1'345	1'9	7079
Dantzick Vitriol.	303 to	200	1'295	1'715	7551
Oil of Vitriol.	10 to	7	1'041	1'7	6124
Rain Water.	529 to	396	0'7845	1'	7845
Gum Arabick.	31 to	21	1'179	1'375	8574
Spirit of Wine well rectified.	100 to	73	0'8765	0'866	10121
Camphire.	3 to	2	1'25	0'996	12551
Oil Olive.	22 to	15	1'1511	0'913	12607
Linseed Oil.	40 to	27	1'1948	0'932	12819
Spirit of Turpentine.	25 to	17	1'1626	0'874	13222
Amber.	14 to	9	1'42	1'04	13654
A Diamond.	100 to	41	4'949	3'4	14556

The Refraction of the Air in this Table is deter-

min'd by that of the Atmosphere observed by Astronomers. For, if Light pass through many refracting
Substances or Mediums gradually denser and denser,
and terminated with parallel Surfaces, the Sum of
all the Refractions will be equal to the single Refraction which it would have suffer'd in passing immediately out of the first Medium into the last. And this
holds true, though the Number of the refracting
Substances be increased to Infinity, and the Distances from one another as much decreased, so that
the Light may be refracted in every Point of its
Passage, and by continual Refractions bent into a
Curve-Line. And therefore the whole Refraction of
Light in passing through the Atmosphere from the
highest and rarest Part thereof down to the lowest
and densest Part, must be equal to the Refraction
which it would suffer in passing at like Obliquity out
of a Vacuum immediately into Air of equal Density
with that in the lowest Part of the Atmosphere.

Now, although a Pseudo-Topaz, a Selenitis, Rock
Crystal, Island Crystal, Vulgar Glass (that is, Sand
melted together) and Glass of Antimony, which are
terrestrial stony alcalizate Concretes, and Air which
probably arises from such Substances by Fermentation, be Substances very differing from one another
in Density, yet by this Table, they have their refractive Powers almost in the same Proportion to one
another as their Densities are, excepting that the Refraction of that strange Substance, Island Crystal is
a little bigger than the rest. And particularly Air,
which is 3500 Times rarer than the Pseudo-Topaz,

and 4400 Times rarer than Glass of Antimony, and 2000 Times rarer than the Selenitis, Glass vulgar, or Crystal of the Rock, has notwithstanding its rarity the same refractive Power in respect of its Density which those very dense Substances have in respect of theirs, excepting so far as those differ from one another.

Again, the Refraction of Camphire, Oil Olive, Linseed Oil, Spirit of Turpentine and Amber, which are fat sulphureous unctuous Bodies, and a Diamond, which probably is an unctuous Substance coagulated, have their refractive Powers in Proportion to one another as their Densities without any considerable Variation. But the refractive Powers of these unctuous Substances are two or three Times greater in respect of their Densities than the refractive Powers of the former Substances in respect of theirs.

Water has a refractive Power in a middle degree between those two sorts of Substances, and probably is of a middle nature. For out of it grow all vegetable and animal Substances, which consist as well of sulphureous fat and inflamable Parts, as of earthy lean and alcalizate ones.

Salts and Vitriols have refractive Powers in a middle degree between those of earthy Substances and Water, and accordingly are composed of those two sorts of Substances. For by distillation and rectification of their Spirits a great Part of them goes into Water, and a great Part remains behind in the form of a dry fix'd Earth capable of Vitrification.

Spirit of Wine has a refractive Power in a middle

degree between those of Water and oily Substances, and accordingly seems to be composed of both, united by Fermentation; the Water, by means of some saline Spirits with which 'tis impregnated, dissolving the Oil, and volatizing it by the Action. For Spirit of Wine is inflamable by means of its oily Parts, and being distilled often from Salt of Tartar, grow by every distillation more and more aqueous and phlegmatick. And Chymists observe, that Vegetables (as Lavender, Rue, Marjoram, &c.) distilled *per se*, before fermentation yield Oils without any burning Spirits, but after fermentation yield ardent Spirits without Oils: Which shews, that their Oil is by fermentation converted into Spirit. They find also, that if Oils be poured in a small quantity upon fermentating Vegetables, they distil over after fermentation in the form of Spirits.

So then, by the foregoing Table, all Bodies seem to have their refractive Powers proportional to their Densities, (or very nearly;) excepting so far as they partake more or less of sulphureous oily Particles, and thereby have their refractive Power made greater or less. Whence it seems rational to attribute the refractive Power of all Bodies chiefly, if not wholly, to the sulphureous Parts with which they abound. For it's probable that all Bodies abound more or less with Sulphurs. And as Light congregated by a Burning-glass acts most upon sulphureous Bodies, to turn them into Fire and Flame ; so, since all Action is mutual, Sulphurs ought to act most upon Light. For that the action between Light and

Bodies is mutual, may appear from this Consideration; That the densest Bodies which refract and reflect Light most strongly, grow hottest in the Summer Sun, by the action of the refracted or reflected
Light.

I have hitherto explain'd the power of Bodies to
reflect and refract, and shew'd, that thin transparent
Plates, Fibres, and Particles, do, according to their
several thicknesses and densities, reflect several sorts
of Rays, and thereby appear of several Colours; and
by consequence that nothing more is requisite for
producing all the Colours of natural Bodies, than the
several sizes and densities of their transparent Particles. But whence it is that these Plates, Fibres, and
Particles, do, according to their several thicknesses
and densities, reflect several sorts of Rays, I have not
yet explain'd. To give some insight into this matter,
and make way for understanding the next part of
this Book, I shall conclude this part with a few more
Propositions. Those which preceded respect the
nature of Bodies, these the nature of Light: For both
must be understood, before the reason of their
Actions upon one another can be known. And because the last Proposition depended upon the velocity of Light, I will begin with a Proposition of that
kind.

PROP. XI.

Light is propagated from luminous Bodies in time, and spends about seven or eight Minutes of an Hour in passing from the Sun to the Earth.

THIS was observed first by *Roemer*, and then by others, by means of the Eclipses of the Satellites of *Jupiter*. For these Eclipses, when the Earth is between the Sun and *Jupiter*, happen about seven or eight Minutes sooner than they ought to do by the Tables, and when the Earth is beyond the Sun they happen about seven or eight Minutes later than they ought to do; the reason being, that the Light of the Satellites has farther to go in the latter case than in the former by the Diameter of the Earth's Orbit. Some inequalities of time may arise from the Excentricities of the Orbs of the Satellites; but those cannot answer in all the Satellites, and at all times to the Position and Distance of the Earth from the Sun. The mean motions of *Jupiter*'s Satellites is also swifter in his descent from his Aphelium to his Perihelium, than in his ascent in the other half of his Orb. But this inequality has no respect to the position of the Earth, and in the three interior Satellites is insensible, as I find by computation from the Theory of their Gravity.

PROP. XII.

*Every Ray of Light in its passage through any refract-
ing Surface is put into a certain transient Constitution or State, which in the progress of the Ray
returns at equal Intervals, and disposes the Ray at
every return to be easily transmitted through the
next refracting Surface, and between the returns
to be easily reflected by it.*

THIS is manifest by the 5th, 9th, 12th, and 15th
Observations. For by those Observations it
appears, that one and the same sort of Rays at equal
Angles of Incidence on any thin transparent Plate,
is alternately reflected and transmitted for many
Successions accordingly as the thickness of the Plate
increases in arithmetical Progression of the Numbers,
0, 1, 2, 3, 4, 5, 6, 7, 8, &c. so that if the first Reflexion (that which makes the first or innermost of
the Rings of Colours there described) be made at the
thickness 1, the Rays shall be transmitted at the
thicknesses 0, 2, 4, 6, 8, 10, 12, &c. and thereby make
the central Spot and Rings of Light, which appear
by transmission, and be reflected at the thickness 1,
3, 5, 7, 9, 11, &c. and thereby make the Rings which
appear by Reflexion. And this alternate Reflexion
and Transmission, as I gather by the 24th Observation, continues for above an hundred vicissitudes,
and by the Observations in the next part of this Book,
for many thousands, being propagated from one Surface of a Glass Plate to the other, though the thick-

ness of the Plate be a quarter of an Inch or above: So that this alternation seems to be propagated from every refracting Surface to all distances without end or limitation.

This alternate Reflexion and Refraction depends on both the Surfaces of every thin Plate, because it depends on their distance. By the 21st Observation, if either Surface of a thin Plate of *Muscovy* Glass be wetted, the Colours caused by the alternate Reflexion and Refraction grow faint, and therefore it depends on them both.

It is therefore perform'd at the second Surface; for if it were perform'd at the first, before the Rays arrive at the second, it would not depend on the second.

It is also influenced by some action or disposition, propagated from the first to the second, because otherwise at the second it would not depend on the first. And this action or disposition, in its propagation, intermits and returns by equal Intervals, because in all its progress it inclines the Ray at one distance from the first Surface to be reflected by the second, at another to be transmitted by it, and that by equal Intervals for innumerable vicissitudes. And because the Ray is disposed to Reflexion at the distances 1, 3, 5, 7, 9, &c. and to Transmission at the distances 0, 2, 4, 6, 8, 10, &c. (for its transmission through the first Surface, is at the distance 0, and it is transmitted through both together, if their distance be infinitely little or much less than 1) the disposition to be transmitted at the distances 2, 4, 6, 8, 10,

&c. is to be accounted a return of the same disposition which the Ray first had at the distance o, that is at its transmission through the first refracting Surface. All which is the thing I would prove.

What kind of action or disposition this is; Whether it consists in a circulating or a vibrating motion of the Ray, or of the Medium, or something else, I do not here enquire. Those that are averse from assenting to any new Discoveries, but such as they can explain by an Hypothesis, may for the present suppose, that as Stones by falling upon Water put the Water into an undulating Motion, and all Bodies by percussion excite vibrations in the Air; so the Rays of Light, by impinging on any refracting or reflecting Surface, excite vibrations in the refracting or reflecting Medium or Substance, and by exciting them agitate the solid parts of the refracting or reflecting Body, and by agitating them cause the Body to grow warm or hot; that the vibrations thus excited are propagated in the refracting or reflecting Medium or Substance, much after the manner that vibrations are propagated in the Air for causing Sound, and move faster than the Rays so as to overtake them; and that when any Ray is in that part of the vibration which conspires with its Motion, it easily breaks through a refracting Surface, but when it is in the contrary part of the vibration which impedes its Motion, it is easily reflected; and, by consequence, that every Ray is successively disposed to be easily reflected, or easily transmitted, by every vibration which overtakes it. But whether this Hypothesis be true or false

I do not here consider. I content my self with the bare Discovery, that the Rays of Light are by some cause or other alternately disposed to be reflected or refracted for many vicissitudes.

DEFINITION.

The returns of the disposition of any Ray to be reflected I will call its Fits of easy Reflexion, *and those of its disposition to be transmitted its* Fits of easy Transmission, *and the space it passes between every return and the next return, the* Interval of its Fits.

PROP. XIII.

The reason why the Surfaces of all thick transparent Bodies reflect part of the Light incident on them, and refract the rest, is, that some Rays at their Incidence are in Fits of easy Reflexion, and others in Fits of easy Transmission.

THIS may be gather'd from the 24th Observation, where the Light reflected by thin Plates of Air and Glass, which to the naked Eye appear'd evenly white all over the Plate, did through a Prism appear waved with many Successions of Light and Darkness made by alternate Fits of easy Reflexion and easy Transmission, the Prism severing and distinguishing the Waves of which the white reflected Light was composed, as was explain'd above.

And hence Light is in Fits of easy Reflexion and easy Transmission, before its Incidence on transparent Bodies. And probably it is put into such fits at its first emission from luminous Bodies, and continues in them during all its progress. For these Fits are of a lasting nature, as will appear by the next part of this Book.

In this Proposition I suppose the transparent Bodies to be thick ; because if the thickness of the Body be much less than the Interval of the Fits of easy Reflexion and Transmission of the Rays, the Body loseth its reflecting power. For if the Rays, which at their entering into the Body are put into Fits of easy Transmission, arrive at the farthest Surface of the Body before they be out of those Fits, they must be transmitted. And this is the reason why Bubbles of Water lose their reflecting power when they grow very thin; and why all opake Bodies, when reduced into very small parts, become transparent.

PROP. XIV.

Those Surfaces of transparent Bodies, which if the Ray be in a Fit of Refraction do refract it most strongly, if the Ray be in a Fit of Reflexion do reflect it most easily.

FOR we shewed above, in *Prop.* 8. that the cause of Reflexion is not the impinging of Light on the solid impervious parts of Bodies, but some other power by which those solid parts act on Light at a distance. We shewed also in *Prop.* 9. that Bodies reflect and refract Light by one and the same power, variously exercised in various circumstances; and in *Prop.* 1. that the most strongly refracting Surfaces reflect the most Light: All which compared together evince and rarify both this and the last Proposition.

PROP. XV.

In any one and the same sort of Rays, emerging in any Angle out of any refracting Surface into one and the same Medium, the Interval of the following Fits of easy Reflexion and Transmission are either accurately or very nearly, as the Rectangle of the Secant of the Angle of Refraction, and of the Secant of another Angle, whose Sine is the first of 106 arithmetical mean Proportionals, between the Sines of Incidence and Refraction, counted from the Sine of Refraction.

THIS is manifest by the 7th and 19th Observations.

Prop. XVI.

In several sorts of Rays emerging in equal Angles out of any refracting Surface into the same Medium, the Intervals of the following Fits of easy Reflexion and easy Transmission are either accurately, or very nearly, as the Cube-Roots of the Squares of the lengths of a Chord, which found the Notes in an Eight, sol, la, fa, sol, la, mi, fa, sol, *with all their imtermediate degrees answering to the Colours of those Rays, according to the Analogy described in the seventh Experiment of the second Part of the first Book.*

THIS is manifest by the 13th and 14th Observations.

Prop. XVII.

If Rays of any sort pass perpendicularly into several Mediums, the Intervals of the Fits of easy Reflexion and Transmission in any one Medium, are to those Intervals in any other, as the Sine of Incidence to the Sine of Refraction, when the Rays pass out of the first of those two Mediums into the second.

THIS is manifest by the 10th Observation.

Prop. XVIII.

*If the Rays which paint the Colour in the Confine of
yellow and orange pass perpendicularly out of any
Medium into Air, the Intervals of their Fits of
easy Reflexion are the $\frac{1}{89000}$ th part of an Inch.
And of the same length are the Intervals of their
Fits of easy Transmission.*

THIS is manifest by the 6th Observation.
From these Propositions it is easy to collect the Intervals of the Fits of easy Reflexion and
easy Transmission of any sort of Rays refracted in
any angle into any Medium; and thence to know,
whether the Rays shall be reflected or transmitted at
their subsequent Incidence upon any other pellucid
Medium. Which thing, being useful for understanding the next part of this Book, was here to be set
down. And for the same reason I add the two following Propositions.

Prop. XIX.

If any sort of Rays falling on the polite Surface of any pellucid Medium be reflected back, the Fits of easy Reflexion, which they have at the point of Reflexion, shall still continue to return; and the Returns shall be at distances from the point of Reflexion in the arithmetical progression of the Numbers 2, 4, 6, 8, 10, 12, &c. and between these Fits the Rays shall be in Fits of easy Transmission.

FOR since the Fits of easy Reflexion and easy Transmission are of a returning nature, there is no reason why these Fits, which continued till the Ray arrived at the reflecting Medium, and there inclined the Ray to Reflexion, should there cease. And if the Ray at the point of Reflexion was in a Fit of easy Reflexion, the progression of the distances of these Fits from that point must begin from 0, and so be of the Numbers 0, 2, 4, 6, 8, &c. And therefore the progression of the distances of the intermediate Fits of easy Transmission, reckon'd from the same point, must be in the progression of the odd Numbers 1, 3, 5, 7, 9, &c. contrary to what happens when the Fits are propagated from points of Refraction.

PROP. XX.

The Intervals of the Fits of easy Reflexion and easy Transmission, propagated from points of Reflexion into any Medium, are equal to the Intervals of the like Fits, which the same Rays would have, if refracted into the same Medium in Angles of Refraction equal to their Angles of Reflexion.

FOR when Light is reflected by the second Surface of thin Plates, it goes out afterwards freely at the first Surface to make the Rings of Colours which appear by Reflexion; and, by the freedom of its egress, makes the Colours of these Rings more vivid and strong than those which appear on the other side of the Plates by the transmitted Light. The reflected Rays are therefore in Fits of easy Transmission at their egress; which would not always happen, if the Intervals of the Fits within the Plate after Reflexion were not equal, both in length and number, to their Intervals before it. And this confirms also the proportions set down in the former Proposition. For if the Rays both in going in and out at the first Surface be in Fits of easy Transmission, and the Intervals and Numbers of those Fits between the first and second Surface, before and after Reflexion, be equal, the distances of the Fits of easy Transmission from either Surface, must be in the same progression after Reflexion as before; that is, from the first Surface which transmitted them in the progression of the even Numbers 0, 2, 4, 6,

8, &c. and from the second which reflected them, in that of the odd Numbers 1, 3, 5, 7, &c. But these two Propositions will become much more evident by the Observations in the following part of this Book.

THE
SECOND BOOK
OF
OPTICKS

PART IV.

Observations concerning the Reflexions and Colours of
thick transparent polish'd Plates.

THERE is no Glass or Speculum how well soever polished, but, besides the Light which
it refracts or reflects regularly, scatters every way
irregularly a faint Light, by means of which the
polish'd Surface, when illuminated in a dark room
by a beam of the Sun's Light, may be easily seen in
all positions of the Eye. There are certain Phænomena of this scatter'd Light, which when I first observed them, seem'd very strange and surprizing to
me. My Observations were as follows.

Obs. 1. The Sun shining into my darken'd
Chamber through a hole one third of an Inch wide,

I let the intromitted beam of Light fall perpendicularly upon a Glass Speculum ground concave on
one side and convex on the other, to a Sphere of five
Feet and eleven Inches Radius, and Quick-silver'd
over on the convex side. And holding a white opake
Chart, or a Quire of Paper at the center of the
Spheres to which the Speculum was ground, that is,
at the distance of about five Feet and eleven Inches
from the Speculum, in such manner, that the beam
of Light might pass through a little hole made in the
middle of the Chart to the Speculum, and thence be
reflected back to the same hole : I observed upon the
Chart four or five concentric Irises or Rings of
Colours, like Rainbows, encompassing the hole much
after the manner that those, which in the fourth
and following Observations of the first part of this
Book appear'd between the Object-glasses, encompassed the black Spot, but yet larger and fainter than
those. These Rings as they grew larger and larger
became diluter and fainter, so that the fifth was scarce
visible. Yet sometimes, when the Sun shone very
clear, there appear'd faint Lineaments of a sixth and
seventh. If the distance of the Chart from the Speculum was much greater or much less than that of six
Feet, the Rings became dilute and vanish'd. And
if the distance of the Speculum from the Window
was much greater than that of six Feet, the reflected
beam of Light would be so broad at the distance
of six Feet from the Speculum where the Rings
appear'd, as to obscure one or two of the innermost
Rings. And therefore I usually placed the Speculum

at about six Feet from the Window; so that its Focus might there fall in with the center of its concavity at the Rings upon the Chart. And this Posture is always to be understood in the following Observations where no other is express'd.

Obs. 2. The Colours of these Rain-bows succeeded one another from the center outwards, in the same form and order with those which were made in the ninth Observation of the first Part of this Book by Light not reflected, but transmitted through the two Object-glasses. For, first, there was in their common center a white round Spot of faint Light, something broader than the reflected beam of Light, which beam sometimes fell upon the middle of the Spot, and sometimes by a little inclination of the Speculum receded from the middle, and left the Spot white to the center.

This white Spot was immediately encompassed with a dark grey or russet, and that dark grey with the Colours of the first Iris; which Colours on the inside next the dark grey were a little violet and indigo, and next to that a blue, which on the outside grew pale, and then succeeded a little greenish yellow, and after that a brighter yellow, and then on the outward edge of the Iris a red which on the outside inclined to purple.

This Iris was immediately encompassed with a second, whose Colours were in order from the inside outwards, purple, blue, green, yellow, light red, a red mix'd with purple.

Then immediately follow'd the Colours of the

third Iris, which were in order outwards a green inclining to purple, a good green, and a red more
bright than that of the former Iris.

The fourth and fifth Iris seem'd of a bluish green
within, and red without, but so faintly that it was
difficult to discern the Colours.

Obs. 3. Measuring the Diameters of these Rings
upon the Chart as accurately as I could, I found
them also in the same proportion to one another
with the Rings made by Light transmitted through
the two Object-glasses. For the Diameters of the
four first of the bright Rings measured between the
brightest parts of their Orbits, at the distance of six
Feet from the Speculum were $1\frac{11}{16}$, $2\frac{3}{8}$, $2\frac{11}{12}$, $3\frac{3}{8}$
Inches, whose Squares are in arithmetical progres-
sion of the numbers 1, 2, 3, 4. If the white circular
Spot in the middle be reckon'd amongst the Rings,
and its central Light, where it seems to be most
luminous, be put equipollent to an infinitely little
Ring; the Squares of the Diameters of the Rings will
be in the progression 0, 1, 2, 3, 4, *&c.* I measured
also the Diameters of the dark Circles between these
luminous ones, and found their Squares in the pro-
gression of the numbers $\frac{1}{2}$, $1\frac{1}{2}$, $2\frac{1}{2}$, $3\frac{1}{2}$, *&c.* the Diameters of the first four at the distance of six Feet
from the Speculum, being $1\frac{3}{16}$, $2\frac{1}{16}$, $2\frac{2}{3}$, $3\frac{3}{20}$ Inches.
If the distance of the Chart from the Speculum was
increased or diminished, the Diameters of the Circles
were increased or diminished proportionally.

Obs. 4. By the analogy between these Rings and
those described in the Observations of the first Part

of this Book, I suspected that there were many more of them which spread into one another, and by interfering mix'd their Colours, and diluted one another so that they could not be seen apart. I viewed them therefore through a Prism, as I did those in the 24th Observation of the first Part of this Book. And when the Prism was so placed as by refracting the Light of their mix'd Colours to separate them, and distinguish the Rings from one another, as it did those in that Observation, I could then see them distincter than before, and easily number eight or nine of them, and sometimes twelve or thirteen. And had not their Light been so very faint, I question not but that I might have seen many more.

Obs. 5. Placing a Prism at the Window to refract the intromitted beam of Light, and cast the oblong Spectrum of Colours on the Speculum: I covered the Speculum with a black Paper which had in the middle of it a hole to let any one of the Colours pass through to the Speculum, whilst the rest were intercepted by the Paper. And now I found Rings of that Colour only which fell upon the Speculum. If the Speculum was illuminated with red, the Rings were totally red with dark Intervals, if with blue they were totally blue, and so of the other Colours. And when they were illuminated with any one Colour, the Squares of their Diameters measured between their most luminous Parts, were in the arithmetical Progression of the Numbers, 0, 1, 2, 3, 4 and the Squares of the Diameters of their dark Intervals in the Progression of the intermediate Numbers $\frac{1}{2}$, $1\frac{1}{2}$, $2\frac{1}{2}$, $3\frac{1}{2}$.

But if the Colour was varied, they varied their Magnitude. In the red they were largest, in the indigo and violet least, and in the intermediate Colours yellow, green, and blue, they were of several intermediate Bignesses answering to the Colour, that is, greater in yellow than in green, and greater in green than in blue. And hence I knew, that when the Speculum was illuminated with white Light, the red and yellow on the outside of the Rings were produced by the least refrangible Rays, and the blue and violet by the most refrangible, and that the Colours of each Ring spread into the Colours of the neighbouring Rings on either side, after the manner explain'd in the first and second Part of this Book, and by mixing diluted one another so that they could not be distinguish'd, unless near the Center where they were least mix'd. For in this Observation I could see the Rings more distinctly, and to a greater Number than before, being able in the yellow Light to number eight or nine of them, besides a faint shadow of a tenth. To satisfy my self how much the Colours of the several Rings spread into one another, I measured the Diameters of the second and third Rings, and found them when made by the Confine of the red and orange to be to the same Diameters when made by the Confine of blue and indigo, as 9 to 8, or thereabouts. For it was hard to determine this Proportion accurately. Also the Circles made successively by the red, yellow, and green, differ'd more from one another than those made successively by the green, blue, and indigo. For the Circle made by the violet was too dark to be seen. To carry on the Computation, let us therefore suppose that the Differences of the Diameters of the Circles made by the outmost red, the Confine of red and orange, the Confine of orange and yellow, the Confine of yellow and green, the Confine of green and blue, the Confine of blue and indigo, the Confine of indigo and violet, and outmost violet, are in proportion as the Differences of the Lengths of a Monochord which sound the Tones in an Eight; *sol, la, fa sol, la, mi, fa, sol*, that is, as the Numbers $\frac{1}{9}, \frac{1}{18}, \frac{1}{12}, \frac{1}{12}, \frac{2}{27}, \frac{1}{27}, \frac{1}{18}$. And if the Diameter of the Circle made by the Confine of red and orange be 9A, and that of the Circle made by the Confine of blue and indigo be 8A as above ; their difference 9A – 8A will be to the difference of the Diameters of the Circles made by the outmost red, and by the Confine of red and orange, as $\frac{1}{18} + \frac{1}{12} + \frac{1}{12} + \frac{2}{27}$ to $\frac{1}{9}$, that is as $\frac{8}{27}$ to $\frac{1}{9}$, or 8 to 3, and to the difference of the Circles made by the outmost violet, and by the Confine of blue and indigo, as $\frac{1}{18} + \frac{1}{12} + \frac{1}{12} + \frac{2}{27}$ to $\frac{1}{27} + \frac{1}{18}$, that is, as $\frac{8}{27}$ to $\frac{5}{54}$, or as 16 to 5. And therefore these differences will be $\frac{3}{8}$A and $\frac{5}{16}$A. Add the first to 9A and subduct the last from 8A, and you will have the Diameters of the Circles made by the least and most refrangible Rays $\frac{75}{8}$A and $\frac{61\frac{1}{2}}{8}$A. These diameters are therefore to one another as 75 to 61½ or 50 to 41, and their Squares as 2500 to 1681, that is, as 3 to 2 very nearly. Which proportion differs not much from the proportion of the Diameters of the Circles made by the outmost red and outmost violet, in the 13th Observation of the first part of this Book.

Obs. 6. Placing my Eye where these Rings appear'd plainest, I saw the Speculum tinged all over with Waves of Colours, (red, yellow, green, blue;) like those which in the Observations of the first part of this Book appeared between the Object-glasses, and upon Bubbles of Water, but much larger. And after the manner of those, they were of various magnitudes in various Positions of the Eye, swelling and shrinking as I moved my Eye this way and that way. They were formed like Arcs of concentrick Circles, as those were; and when my Eye was over against the center of the concavity of the Speculum, (that is, 5 Feet and 10 Inches distant from the Speculum,) their common center was in a right Line with that center of concavity, and with the hole in the Window. But in other postures of my Eye their center had other positions. They appear'd by the Light of the Clouds propagated to the Speculum through the hole in the Window; and when the Sun shone through that hole upon the Speculum, his Light upon it was of the Colour of the Ring whereon it fell, but by its splendor obscured the Rings made by the Light of the Clouds, unless when the Speculum was removed to a great distance from the Window, so that his Light upon it might be broad and faint. By varying the position of my Eye, and moving it nearer to or farther from the direct beam of the Sun's Light, the Colour of the Sun's reflected Light constantly varied upon the Speculum, as it did upon my

Eye, the same Colour always appearing to a By-
stander upon my Eye which to me appear'd upon the
Speculum. And thence I knew that the Rings of
Colours upon the Chart were made by these reflected Colours, propagated thither from the Speculum in several Angles, and that their production
depended not upon the termination of Light and
Shadow.

Obs. 7. By the Analogy of all these Phænomena
with those of the like Rings of Colours described in
the first part of this Book, it seemed to me that these
Colours were produced by this thick Plate of Glass,
much after the manner that those were produced by
very thin Plates. For, upon trial, I found that if the
Quick-silver were rubb'd off from the backside of the
Speculum, the Glass alone would cause the same
Rings of Colours, but much more faint than before;
and therefore the Phænomenon depends not upon
the Quick-silver, unless so far as the Quick-silver
by increasing the Reflexion of the backside of the
Glass increases the Light of the Rings of Colours.
I found also that a Speculum of Metal without Glass
made some Years since for optical uses, and very well
wrought, produced none of those Rings; and thence
I understood that these Rings arise not from one
specular Surface alone, but depend upon the two
Surfaces of the Plate of Glass whereof the Speculum
was made, and upon the thickness of the Glass between them. For as in the 7th and 19th Observations
of the first part of this Book a thin Plate of Air,
Water, or Glass of an even thickness appeared of one

Colour when the Rays were perpendicular to it, of another when they were a little oblique, of another when more oblique, of another when still more oblique, and so on; so here, in the sixth Observation, the Light which emerged out of the Glass in several Obliquities, made the Glass appear of several Colours, and being propagated in those Obliquities to the Chart, there painted Rings of those Colours. And as the reason why a thin Plate appeared of several Colours in several Obliquities of the Rays, was, that the Rays of one and the same sort are reflected by the thin Plate at one obliquity and transmitted at another, and those of other sorts transmitted where these are reflected, and reflected where these are transmitted: So the reason why the thick Plate of Glass whereof the Speculum was made did appear of various Colours in various Obliquities, and in those Obliquities propagated those Colours to the Chart, was, that the Rays of one and the same sort did at one Obliquity emerge out of the Glass, at another did not emerge, but were reflected back towards the Quick-silver by the hither Surface of the Glass, and accordingly as the Obliquity became greater and greater, emerged and were reflected alternately for many Successions; and that in one and the same Obliquity the Rays of one sort were reflected, and those of another transmitted. This is manifest by the fifth Observation of this part of this Book. For in that Observation, when the Speculum was illuminated by any one of the prismatick Colours, that Light made many Rings of the same Colour upon the Chart with dark Intervals, and therefore at its emergence out of the Speculum was alternately transmitted and not transmitted from the Speculum to the Chart for many Successions, according to the various Obliquities of its Emergence. And when the Colour cast on the Speculum by the Prism was varied, the Rings became of the Colour cast on it, and varied their bigness with their Colour, and therefore the Light was now alternately transmitted and not transmitted from the Speculum to the Chart at other Obliquities than before. It seemed to me therefore that these Rings were of one and the same original with those of thin Plates, but yet with this difference, that those of thin Plates are made by the alternate Reflexions and Transmissions of the Rays at the second Surface of the Plate, after one passage through it; but here the Rays go twice through the Plate before they are alternately reflected and transmitted. First, they go through it from the first Surface to the Quick-silver, and then return through it from the Quick-silver to the first Surface, and there are either transmitted to the Chart or reflected back to the Quick-silver, accordingly as they are in their Fits of easy Reflexion or Transmission when they arrive at that Surface. For the Intervals of the Fits of the Rays which fall perpendicularly on the Speculum, and are reflected back in the same perpendicular Lines, by reason of the equality of these Angles and Lines, are of the same length and number within the Glass after Reflexion as before, by the 19th Proposition of the third part of this Book.

And therefore since all the Rays that enter through the first Surface are in their Fits of easy Transmission at their entrance, and as many of these as are reflected by the second are in their Fits of easy Reflexion there, all these must be again in their Fits of easy Transmission at their return to the first, and by consequence there go out of the Glass to the Chart, and form upon it the white Spot of Light in the center of the Rings. For the reason holds good in all sorts of Rays, and therefore all sorts must go out promiscuously to that Spot, and by their mixture cause it to be white. But the Intervals of the Fits of those Rays which are reflected more obliquely than they enter, must be greater after Reflexion than before, by the 15th and 20th Propositions. And thence it may happen that the Rays at their return to the first Surface, may in certain Obliquities be in Fits of easy Reflexion, and return back to the Quicksilver, and in other intermediate Obliquities be again in Fits of easy Transmission, and so go out to the Chart, and paint on it the Rings of Colours about the white Spot. And because the Intervals of the Fits at equal obliquities are greater and fewer in the less refrangible Rays, and less and more numerous in the more refrangible, therefore the less refrangible at equal obliquities shall make fewer Rings than the more refrangible, and the Rings made by those shall be larger than the like number of Rings made by these; that is, the red Rings shall be larger than the yellow, the yellow than the green, the green than the blue, and the blue than the violet, as they were really

found to be in the fifth Observation. And therefore
the first Ring of all Colours encompassing the white
Spot of Light shall be red without any violet within,
and yellow, and green, and blue in the middle, as it
was found in the second Observation; and these
Colours in the second Ring, and those that follow,
shall be more expanded, till they spread into one
another, and blend one another by interfering.

These seem to be the reasons of these Rings in
general; and this put me upon observing the thickness of the Glass, and considering whether the
dimensions and proportions of the Rings may be
truly derived from it by computation.

Obs. 8. I measured therefore the thickness of this
concavo-convex Plate of Glass, and found it every
where $\frac{1}{4}$ of an Inch precisely. Now, by the sixth Observation of the first Part of this Book, a thin Plate of
Air transmits the brightest Light of the first Ring,
that is, the bright yellow, when its thickness is the
$\frac{1}{89000}$th part of an Inch; and by the tenth Observation of the same Part, a thin Plate of Glass transmits
the same Light of the same Ring, when its thickness
is less in proportion of the Sine of Refraction to the
Sine of Incidence, that is, when its thickness is the
$\frac{11}{1513000}$th or $\frac{1}{137545}$th part of an Inch, supposing
the Sines are as 11 to 17. And if this thickness be
doubled, it transmits the same bright Light of the
second Ring; if tripled, it transmits that of the
third, and so on; the bright yellow Light in all these
cases being in its Fits of Transmission.   And therefore if its thickness be multiplied 34386 times, so as

to become $\frac{1}{4}$ of an Inch, it transmits the same bright Light of the 34386th Ring. Suppose this be the bright yellow Light transmitted perpendicularly from the reflecting convex side of the Glass through the concave side to the white Spot in the center of the Rings of Colours on the Chart: And by a Rule in the 7th and 19th Observations in the first Part of this Book, and by the 15th and 20th Propositions of the third Part of this Book, if the Rays be made oblique to the Glass, the thickness of the Glass requisite to transmit the same bright Light of the same Ring in any obliquity, is to this thickness of $\frac{1}{4}$ of an Inch, as the Secant of a certain Angle to the Radius, the Sine of which Angle is the first of an hundred and six arithmetical Means between the Sines of Incidence and Refraction, counted from the Sine of Incidence when the Refraction is made out of any plated Body into any Medium encompassing it; that is, in this case, out of Glass into Air. Now if the thickness of the Glass be increased by degrees, so as to bear to its first thickness, (*viz.* that of a quarter of an Inch,) the Proportions which 34386 (the number of Fits of the perpendicular Rays in going through the Glass towards the white Spot in the center of the Rings,) hath to 34385, 34384, 34383, and 34382, (the numbers of the Fits of the oblique Rays in going through the Glass towards the first, second, third, and fourth Rings of Colours,) and if the first thickness be divided into 100000000 equal parts, the increased thicknesses will be 100002908, 100005816, 100008725, and 100011633, and the Angles of which these thicknesses are Secants will be $26' 13''$, $37' 5''$, $45' 6''$, and $52' 26''$, the Radius being 100000000; and the Sines of these Angles are 762, 1079, 1321, and 1525, and the proportional Sines of Refraction 1172, 1659, 2031, and 2345, the Radius being 100000. For since the Sines of Incidence out of Glass into Air are to the Sines of Refraction as 11 to 17, and to the above-mentioned Secants as 11 to the first of 106 arithmetical Means between 11 and 17, that is, as 11 to $11 \frac{6}{106}$, those Secants will be to the Sines of Refraction as $11 \frac{6}{106}$, to 17, and by this Analogy will give these Sines. So then, if the obliquities of the Rays to the concave Surface of the Glass be such that the Sines of their Refraction in passing out of the Glass through that Surface into the Air be 1172, 1659, 2031, 2345, the bright Light of the 34386th Ring shall emerge at the thicknesses of the Glass, which are to $\frac{1}{4}$ of an Inch as 34386 to 34385, 34384, 34383, 34382, respectively. And therefore, if the thickness in all these Cases be $\frac{1}{4}$ of an Inch (as it is in the Glass of which the Speculum was made) the bright Light of the 34385th Ring shall emerge where the Sine of Refraction is 1172, and that of the 34384th, 34383th, and 34382th Ring where the Sine is 1659, 2031, and 2345 respectively. And in these Angles of Refraction the Light of these Rings shall be propagated from the Speculum to the Chart, and there paint Rings about the white central round Spot of Light which we said was the Light of the 34386th Ring. And the Semidiameters of these Rings shall subtend the Angles of Refraction made at the Con-

cave-Surface of the Speculum, and by consequence
their Diameters shall be to the distance of the Chart
from the Speculum as those Sines of Refraction
doubled are to the Radius, that is, as 1172, 1659,
2031, and 2345, doubled are to 100000. And therefore, if the distance of the Chart from the Concave-
Surface of the Speculum be six Feet (as it was in the
third of these Observations) the Diameters of the
Rings of this bright yellow Light upon the Chart
shall be 1'688, 2'389, 2'925, 3'375 Inches: For these
Diameters are to six Feet, as the above-mention'd
Sines doubled are to the Radius. Now, these Diameters of the bright yellow Rings, thus found by
Computation are the very same with those found in
the third of these Observations by measuring them,
viz. with $1\frac{11}{16}$, $2\frac{3}{8}$, $2\frac{11}{12}$, and $3\frac{3}{8}$ Inches, and therefore
the Theory of deriving these Rings from the thick-
ness of the Plate of Glass of which the Speculum was
made, and from the Obliquity of the emerging Rays
agrees with the Observation. In this Computation
I have equalled the Diameters of the bright Rings
made by Light of all Colours, to the Diameters of the
Rings made by the bright yellow. For this yellow
makes the brightest Part of the Rings of all Colours.
If you desire the Diameters of the Rings made by the
Light of any other unmix'd Colour, you may find
them readily by putting them to the Diameters of
the bright yellow ones in a subduplicate Proportion
of the Intervals of the Fits of the Rays of those
Colours when equally inclined to the refracting or
reflecting Surface which caused those Fits, that is,

by putting the Diameters of the Rings made by the
Rays in the Extremities and Limits of the seven
Colours, red, orange, yellow, green, blue, indigo,
violet, proportional to the Cube-roots of the Num-
bers, $1, \frac{8}{9}, \frac{5}{6}, \frac{3}{4}, \frac{2}{3}, \frac{3}{5}, \frac{9}{16}, \frac{1}{2}$, which express the Lengths
of a Monochord sounding the Notes in an Eighth :
For by this means the Diameters of the Rings of
these Colours will be found pretty nearly in the same
Proportion to one another, which they ought to have
by the fifth of these Observations.

And thus I satisfy'd my self, that these Rings were
of the same kind and Original with those of thin
Plates, and by consequence that the Fits or alternate
Dispositions of the Rays to be reflected and transmitted are propagated to great distances from every
reflecting and refracting Surface. But yet to put the
matter out of doubt, I added the following Observation.

Obs. 9. If these Rings thus depend on the thickness of the Plate of Glass, their Diameters at equal
distances from several Speculums made of such concavo-convex Plates of Glass as are ground on the
same Sphere, ought to be reciprocally in a subduplicate Proportion of the thicknesses of the Plates of
Glass. And if this Proportion be found true by experience it will amount to a demonstration that these
Rings (like those formed in thin Plates) do depend
on the thickness of the Glass. I procured therefore
another concavo-convex Plate of Glass ground on
both sides to the same Sphere with the former Plate.
Its thickness was $\frac{5}{62}$ Parts of an Inch; and the Dia-

meters of the three first bright Rings measured between the brightest Parts of their Orbits at the distance of six Feet from the Glass were 3. $4\frac{1}{6}$. $5\frac{1}{8}$. Inches. Now, the thickness of the other Glass being $\frac{1}{4}$ of an Inch was to the thickness of this Glass as $\frac{1}{4}$ to $\frac{5}{62}$, that is as 31 to 10, or 310000000 to 100000000, and the Roots of these Numbers are 17607 and 10000, and in the Proportion of the first of these Roots to the second are the Diameters of the bright Rings made in this Observation by the thinner Glass, 3. $4\frac{1}{6}$. $5\frac{1}{8}$, to the Diameters of the same Rings made in the third of these Observations by the thicker Glass $1\frac{11}{16}$, $2\frac{3}{8}$. $2\frac{11}{12}$, that is, the Diameters of the Rings are reciprocally in a subduplicate Proportion of the thicknesses of the Plates of Glass.

So then in Plates of Glass which are alike concave on one side, and alike convex on the other side, and alike quick-silver'd on the convex sides, and differ in nothing but their thickness, the Diameters of the Rings are reciprocally in a subduplicate Proportion of the thicknesses of the Plates. And this shews sufficiently that the Rings depend on both the Surfaces of the Glass. They depend on the convex Surface, because they are more luminous when that Surface is quick-silver'd over than when it is without Quick-silver. They depend also upon the concave Surface, because without that Surface a Speculum makes them not. They depend on both Surfaces, and on the distances between them, because their bigness is varied by varying only that distance. And this dependence is of the same kind with that which the Colours

of thin Plates have on the distance of the Surfaces
of those Plates, because the bigness of the Rings,
and their Proportion to one another, and the variation
of their bigness arising from the variation of the thickness of the Glass, and the Orders of their Colours, is
such as ought to result from the Propositions in the
end of the third Part of this Book, derived from
the Phænomena of the Colours of thin Plates set
down in the first Part.

There are yet other Phænomena of these Rings of
Colours, but such as follow from the same Proposi-
tions, and therefore confirm both the Truth of those
Propositions, and the Analogy between these Rings
and the Rings of Colours made by very thin Plates.
I shall subjoin some of them.

Obs. 10. When the beam of the Sun's Light was
reflected back from the Speculum not directly to the
hole in the Window, but to a place a little distant from
it, the common center of that Spot, and of all the
Rings of Colours fell in the middle way between the
beam of the incident Light, and the beam of the reflected Light, and by consequence in the center of
the spherical concavity of the Speculum, whenever
the Chart on which the Rings of Colours fell was
placed at that center. And as the beam of reflected
Light by inclining the Speculum receded more and
more from the beam of incident Light and from the
common center of the colour'd Rings between them,
those Rings grew bigger and bigger, and so also did
the white round Spot, and new Rings of Colours
emerged successively out of their common center,

and the white Spot became a white Ring encompassing them; and the incident and reflected beams
of Light always fell upon the opposite parts of this
white Ring, illuminating its Perimeter like two mock
Suns in the opposite parts of an Iris. So then the
Diameter of this Ring, measured from the middle
of its Light on one side to the middle of its Light on
the other side, was always equal to the distance between the middle of the incident beam of Light, and
the middle of the reflected beam measured at the
Chart on which the Rings appeared: And the Rays
which form'd this Ring were reflected by the Speculum in Angles equal to their Angles of Incidence,
and by consequence to their Angles of Refraction at
their entrance into the Glass, but yet their Angles of
Reflexion were not in the same Planes with their
Angles of Incidence.

Obs. 11. The Colours of the new Rings were in a
contrary order to those of the former, and arose after
this manner. The white round Spot of Light in the
middle of the Rings continued white to the center
till the distance of the incident and reflected beams
at the Chart was about $\frac{7}{8}$ parts of an Inch, and then it
began to grow dark in the middle. And when that
distance was about $1\frac{3}{16}$ of an Inch, the white Spot
was become a Ring encompassing a dark round Spot
which in the middle inclined to violet and indigo.
And the luminous Rings encompassing it were
grown equal to those dark ones which in the four
first Observations encompassed them, that is to say,
the white Spot was grown a white Ring equal to the

first of those dark Rings, and the first of those luminous Rings was now grown equal to the second of
those dark ones, and the second of those luminous
ones to the third of those dark ones, and so on. For
the Diameters of the luminous Rings were now $1\frac{3}{16}$,
$2\frac{1}{16}$ $2\frac{2}{3}$, $3\frac{3}{20}$, &c. Inches.

When the distance between the incident and reflected beams of Light became a little bigger, there
emerged out of the middle of the dark Spot after the
indigo a blue, and then out of that blue a pale green,
and soon after a yellow and red. And when the
Colour at the center was brightest, being between
yellow and red, the bright Rings were grown equal
to those Rings which in the four first Observations
next encompassed them; that is to say, the white Spot
in the middle of those Rings was now become a white
Ring equal to the first of those bright Rings, and the
first of those bright ones was now become equal to
the second of those, and so on. For the Diameters of
the white Ring, and of the other luminous Rings en-
compassing it, were now $1\frac{11}{16}$, $2\frac{3}{8}$, $2\frac{11}{12}$, $3\frac{3}{8}$, &c. or
thereabouts.

When the distance of the two beams of Light at
the Chart was a little more increased, there emerged
out of the middle in order after the red, a purple, a
blue, a green, a yellow, and a red inclining much to
purple, and when the Colour was brightest being
between yellow and red, the former indigo, blue,
green, yellow and red, were become an Iris or Ring
of Colours equal to the first of those luminous Rings
which appeared in the four first Observations, and

the white Ring which was now become the second of the luminous Rings was grown equal to the second of those, and the first of those which was now become the third Ring was become equal to the third of those, and so on. For their Diameters were $1\frac{11}{16}$, $2\frac{3}{8}$, $2\frac{11}{12}$, $3\frac{3}{8}$ Inches, the distance of the two beams of Light, and the Diameter of the white Ring being $2\frac{3}{8}$ Inches.

When these two beams became more distant there emerged out of the middle of the purplish red, first a darker round Spot, and then out of the middle of that Spot a brighter. And now the former Colours (purple, blue, green, yellow, and purplish red) were become a Ring equal to the first of the bright Rings mentioned in the four first Observations, and the Rings about this Ring were grown equal to the Rings about that respectively; the distance between the two beams of Light and the Diameter of the white Ring (which was now become the third Ring) being about 3 Inches.

The Colours of the Rings in the middle began now to grow very dilute, and if the distance between the two Beams was increased half an Inch, or an Inch more, they vanish'd whilst the white Ring, with one or two of the Rings next it on either side, continued still visible. But if the distance of the two beams of Light was still more increased, these also vanished: For the Light which coming from several parts of the hole in the Window fell upon the Speculum in several Angles of Incidence, made Rings of several bignesses, which diluted and blotted out one another, as I knew by intercepting some part of

that Light. For if I intercepted that part which was
nearest to the Axis of the Speculum the Rings would
be less, if the other part which was remotest from it
they would be bigger.

Obs. 12. When the Colours of the Prism were cast
successively on the Speculum, that Ring which in
the two last Observations was white, was of the same
bigness in all the Colours, but the Rings without it
were greater in the green than in the blue, and still
greater in the yellow, and greatest in the red. And,
on the contrary, the Rings within that white Circle
were less in the green than in the blue, and still less
in the yellow, and least in the red. For the Angles of
Reflexion of those Rays which made this Ring, being
equal to their Angles of Incidence, the Fits of every
reflected Ray within the Glass after Reflexion are
equal in length and number to the Fits of the same
Ray within the Glass before its Incidence on the
reflecting Surface. And therefore since all the Rays
of all sorts at their entrance into the Glass were in a
Fit of Transmission, they were also in a Fit of Transmission at their returning to the same Surface after
Reflexion; and by consequence were transmitted,
and went out to the white Ring on the Chart. This is
the reason why that Ring was of the same bigness in
all the Colours, and why in a mixture of all it appears
white. But in Rays which are reflected in other
Angles, the Intervals of the Fits of the least refrangible being greatest, make the Rings of their Colour
in their progress from this white Ring, either outwards or inwards, increase or decrease by the

greatest steps; so that the Rings of this Colour
without are greatest, and within least. And this
is the reason why in the last Observation, when the
Speculum was illuminated with white Light, the
exterior Rings made by all Colours appeared red
without and blue within, and the interior blue without and red within.

These are the Phænomena of thick convexo-
concave Plates of Glass, which are every where of
the same thickness. There are yet other Phænomena
when these Plates are a little thicker on one side than
on the other, and others when the Plates are more
or less concave than convex, or plano-convex, or
double-convex. For in all these cases the Plates make
Rings of Colours, but after various manners; all
which, so far as I have yet observed, follow from the
Propositions in the end of the third part of this Book,
and so conspire to confirm the truth of those Propositions. But the Phænomena are too various, and
the Calculations whereby they follow from those Propositions too intricate to be here prosecuted. I content my self with having prosecuted this kind of
Phænomena so far as to discover their Cause, and by
discovering it to ratify the Propositions in the third
Part of this Book.

Obs. 13. As Light reflected by a Lens quick-
silver'd on the backside makes the Rings of Colours
above described, so it ought to make the like Rings
of Colours in passing through a drop of Water. At
the first Reflexion of the Rays within the drop, some
Colours ought to be transmitted, as in the case of a

Lens, and others to be reflected back to the Eye. For instance, if the Diameter of a small drop or globule of Water be about the 500th part of an Inch, so that a red-making Ray in passing through the middle of this globule has 250 Fits of easy Transmission within the globule, and that all the red-making Rays which are at a certain distance from this middle Ray round about it have 249 Fits within the globule, and all the like Rays at a certain farther distance round about it have 248 Fits, and all those at a certain farther distance 247 Fits, and so on; these concentrick Circles of Rays after their transmission, falling on a white Paper, will make concentrick Rings of red upon the Paper, supposing the Light which passes through one single globule, strong enough to be sensible. And, in like manner, the Rays of other Colours will make Rings of other Colours. Suppose now that in a fair Day the Sun shines through a thin Cloud of such globules of Water or Hail, and that the globules are all of the same bigness; and the Sun seen through this Cloud shall appear encompassed with the like concentrick Rings of Colours, and the Diameter of the first Ring of red shall be $7\frac{1}{4}$ Degrees, that of the second $10\frac{1}{4}$ Degrees, that of the third 12 Degrees 33 Minutes. And accordingly as the Globules of Water are bigger or less, the Rings shall be less or bigger. This is the Theory, and Experience answers it. For in *June* 1692, I saw by reflexion in a Vessel of stagnating Water three Halos, Crowns, or Rings of Colours about the Sun, like three little Rainbows, concentrick to his Body. The Colours of the first or innermost Crown were blue next the Sun, red without, and white in the middle between the blue and red. Those of the second Crown were purple and blue within, and pale red without, and green in the middle. And those of the third were pale blue within, and pale red without; these Crowns enclosed one another immediately, so that their Colours proceeded in this continual order from the Sun outward: blue, white, red; purple, blue, green, pale yellow and red; pale blue, pale red. The Diameter of the second Crown measured from the middle of the yellow and red on one side of the Sun, to the middle of the same Colour on the other side was $9\frac{1}{3}$ Degrees, or thereabouts. The Diameters of the first and third I had not time to measure, but that of the first seemed to be about five or six Degrees, and that of the third about twelve. The like Crowns appear sometimes about the Moon; for in the beginning of the Year 1664, *Febr.* 19th at Night, I saw two such Crowns about her. The Diameter of the first or innermost was about three Degrees, and that of the second about five Degrees and an half. Next about the Moon was a Circle of white, and next about that the inner Crown, which was of a bluish green within next the white, and of a yellow and red without, and next about these Colours were blue and green on the inside of the outward Crown, and red on the outside of it. At the same time there appear'd a Halo about 22 Degrees 35' distant from the center of the Moon. It was elliptical, and its long Diameter was perpendicular to the Horizon, verging below farthest from the Moon. I am told that the Moon has sometimes three or more concentrick Crowns of Colours encompassing one another next about her Body. The more equal the globules of Water or Ice are to one another, the more Crowns of Colours will appear, and the Colours will be the more lively. The Halo at the distance of 22½ Degrees from the Moon is of another sort. By its being oval and remoter from the Moon below than above, I conclude, that it was made by Refraction in some sort of Hail or Snow floating in the Air in an horizontal posture, the refracting Angle being about 58 or 60 Degrees.

THE
THIRD BOOK
OF
OPTICKS

PART I.

Observations concerning the Inflexions of the Rays of Light, and the Colours made thereby.

GRIMALDO has inform'd us, that if a beam of the Sun's Light be let into a dark Room through a very small hole, the Shadows of things in this Light will be larger than they ought to be if the Rays went on by the Bodies in straight Lines, and that these Shadows have three parallel Fringes, Bands or Ranks of colour'd Light adjacent to them. But if the Hole be enlarged the Fringes grow broad and run into one another, so that they cannot be distinguish'd. These broad Shadows and Fringes have been reckon'd by some to proceed from the ordinary refraction of the Air, but without due examination of the Matter. For the circumstances of the Phænomenon, so far as I have observed them, are as follows.

Obs. 1. I made in a piece of Lead a small Hole with a Pin, whose breadth was the 42d part of an Inch. For 21 of those Pins laid together took up the breadth of half an Inch. Through this Hole I let into my darken'd Chamber a beam of the Sun's Light, and found that the Shadows of Hairs, Thred, Pins, Straws, and such like slender Substances placed in this beam of Light, were considerably broader than they ought to be, if the Rays of Light passed on by these Bodies in right Lines. And particularly a Hair of a Man's Head, whose breadth was but the 280th part of an Inch, being held in this Light, at the distance of about twelve Feet from the Hole, did cast a Shadow which at the distance of four Inches from the Hair was the sixtieth part of an Inch broad, that is, above four times broader than the Hair, and at the distance of two Feet from the Hair was about the eight and twentieth part of an Inch broad, that is, ten times broader than the Hair, and at the distance of ten Feet was the eighth part of an Inch broad, that is 35 times broader.

Nor is it material whether the Hair be encompassed with Air, or with any other pellucid Substance. For I wetted a polish'd Plate of Glass, and laid the Hair in the Water upon the Glass, and then laying another polish'd Plate of Glass upon it, so that the Water might fill up the space between the Glasses, I held them in the aforesaid beam of Light, so that the Light might pass through them perpendicularly, and the Shadow of the Hair was at the same distances as big as before. The Shadows of

Scratches made in polish'd Plates of Glass were also much broader than they ought to be, and the Veins in polish'd Plates of Glass did also cast the like broad Shadows. And therefore the great breadth of these Shadows proceeds from some other cause than the Refraction of the Air. .

Let the Circle X [in *Fig.* 1.] represent the middle of the Hair; ADG, BEH, CFI, three Rays passing by one side of the Hair at several distances; KNQ, LOR, MPS, three other Rays passing by the other side of the Hair at the like distances; D, E, F, and N, O, P, the places where the Rays are bent in their passage by the Hair; G, H, I, and Q, R, S, the places where the Rays fall on a Paper GQ; IS the breadth of the Shadow of the Hair cast on the Paper, and TI, VS, two Rays passing to the Points I and S without bending when the Hair is taken away. And it's manifest that all the Light between these two Rays TI and VS is bent in passing by the Hair, and turned aside from the Shadow IS, because if any part of this Light were not bent it would fall on the Paper within the Shadow, and there illuminate the Paper, contrary to experience. And because when the Paper is at a great distance from the Hair, the Shadow is broad, and therefore the Rays TI and VS are at a great distance from one another, it follows that the Hair acts upon the Rays of Light at a good distance in their passing by it. But the Action is strongest on the Rays which pass by at least distances, and grows weaker and weaker accordingly as the Rays pass by at distances greater and greater, as is represented in the Scheme: For thence it comes to pass, that the Shadow of the Hair is much broader in proportion to the distance of the Paper from the Hair, when the Paper is nearer the Hair, than when it is at a great distance from it.

Obs. 2. The Shadows of all Bodies (Metals, Stones, Glass, Wood, Horn, Ice, &c.) in this Light were border'd with three Parallel Fringes or Bands of colour'd Light, whereof that which was contiguous to the Shadow was broadest and most luminous, and that which was remotest from it was narrowest, and so faint, as not easily to be visible. It was difficult to distinguish the Colours, unless when the Light fell very obliquely upon a smooth Paper, or some other smooth white Body, so as to make them appear much broader than they would otherwise do. And then the Colours were plainly visible in this Order: The first or innermost Fringe was violet and deep blue next the Shadow, and then light blue, green, and yellow in the middle, and red without. The second Fringe was almost contiguous to the first, and the third to the second, and both were blue within, and yellow and red without, but their Colours were very faint, especially those of the third. The Colours therefore proceeded in this order from the Shadow; violet, indigo, pale blue, green, yellow, red; blue, yellow, red; pale blue, pale yellow and red. The Shadows made by Scratches and Bubbles in polish'd Plates of Glass were border'd with the like Fringes of colour'd Light. And if Plates of Looking-glass sloop'd off near the edges with a Diamond-cut, be held in the

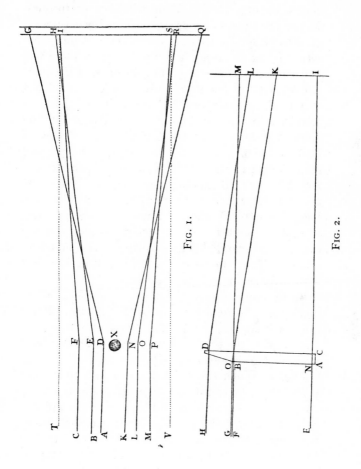

Fig. 1.

Fig. 2.

same beam of Light, the Light which passes through the parallel Planes of the Glass will be border'd with the like Fringes of Colours where those Planes meet with the Diamond-cut, and by this means there will sometimes appear four or five Fringes of Colours. Let AB, CD [in *Fig.* 2.] represent the parallel Planes of a Looking-glass, and BD the Plane of the Diamond-cut, making at B a very obtuse Angle with the Plane AB. And let all the Light between the Rays ENI and FBM pass directly through the parallel Planes of the Glass, and fall upon the Paper between I and M, and all the Light between the Rays GO and HD be refracted by the oblique Plane of the Diamond-cut BD, and fall upon the Paper between K and L; and the Light which passes directly through the parallel Planes of the Glass, and falls upon the Paper between I and M, will be border'd with three or more Fringes at M.

So by looking on the Sun through a Feather or black Ribband held close to the Eye, several Rainbows will appear; the Shadows which the Fibres or Threds cast on the *Tunica Retina*, being border'd with the like Fringes of Colours.

Obs. 3. When the Hair was twelve Feet distant from this Hole, and its Shadow fell obliquely upon a flat white Scale of Inches and Parts of an Inch placed half a Foot beyond it, and also when the Shadow fell perpendicularly upon the same Scale placed nine Feet beyond it; I measured the breadth of the Shadow and Fringes as accurately as I could, and found them in Parts of an Inch as follows.

At the Distance of	half a Foot	Nine Feet
The breadth of the Shadow	$\frac{1}{54}$	$\frac{1}{9}$
The breadth between the Middles of the brightest Light of the inner-most Fringes on either side the Shadow	$\frac{1}{38}$ or $\frac{1}{39}$	$\frac{7}{50}$
The breadth between the Middles of the brightest Light of the middle-most Fringes on either side the Shadow	$\frac{1}{23\frac{1}{2}}$	$\frac{4}{17}$
The breadth between the Middles of the brightest Light of the out-most Fringes on either side the Shadow	$\frac{1}{18}$ or $\frac{1}{18\frac{1}{2}}$	$\frac{3}{10}$
The distance between the Middles of the brightest Light of the first and second Fringes	$\frac{1}{120}$	$\frac{1}{21}$
The distance between the Middles of the brightest Light of the second and third Fringes	$\frac{1}{170}$	$\frac{1}{31}$
The breadth of the luminous Part (green, white, yellow, and red) of the first Fringe	$\frac{1}{170}$	$\frac{1}{32}$
The breadth of the darker Space between the first and second Fringes	$\frac{1}{240}$	$\frac{1}{45}$
The breadth of the luminous Part of the second Fringe	$\frac{1}{290}$	$\frac{1}{55}$
The breadth of the darker Space between the second and third Fringes	$\frac{1}{340}$	$\frac{1}{63}$

These Measures I took by letting the Shadow of the Hair, at half a Foot distance, fall so obliquely on the Scale, as to appear twelve times broader than when it fell perpendicularly on it at the same distance, and setting down in this Table the twelfth part of the Measures I then took.

Obs. 4. When the Shadow and Fringes were cast obliquely upon a smooth white Body, and that Body was removed farther and farther from the Hair, the first Fringe began to appear and look brighter than the rest of the Light at the distance of less than a quarter of an Inch from the Hair, and the dark Line or Shadow between that and the second Fringe began to appear at a less distance from the Hair than that of the third part of an Inch. The second Fringe began to appear at a distance from the Hair of less than half an Inch, and the Shadow between that and the third Fringe at a distance less than an inch, and the third Fringe at a distance less than three Inches. At greater distances they became much more sensible, but kept very nearly the same proportion of their breadths and intervals which they had at their first appearing. For the distance between the middle of the first, and middle of the second Fringe, was to the distance between the middle of the second and middle of the third Fringe, as three to two, or ten to seven. And the last of these two distances was equal to the breadth of the bright Light or luminous part of the first Fringe. And this breadth was to the breadth of the bright Light of the second Fringe as seven to four, and to the dark Interval of the first

and second Fringe as three to two, and to the like
dark Interval between the second and third as two
to one. For the breadths of the Fringes seem'd to be
in the progression of the Numbers 1, $\sqrt{\frac{1}{3}}$, $\sqrt{\frac{1}{5}}$, and
their Intervals to be in the same progression with
them; that is, the Fringes and their Intervals together to be in the continual progression of the
Numbers 1, $\sqrt{\frac{1}{2}}$, $\sqrt{\frac{1}{3}}$, $\sqrt{\frac{1}{4}}$, $\sqrt{\frac{1}{5}}$, or thereabouts. And
these Proportions held the same very nearly at all
distances from the Hair; the dark Intervals of the
Fringes being as broad in proportion to the breadth
of the Fringes at their first appearance as afterwards
at great distances from the Hair, though not so dark
and distinct.

Obs. 5. The Sun shining into my darken'd
Chamber through a hole a quarter of an Inch broad,
I placed at the distance of two or three Feet from
the Hole a Sheet of Pasteboard, which was black'd
all over on both sides, and in the middle of it had a
hole about three quarters of an Inch square for the
Light to pass through. And behind the hole I
fasten'd to the Pasteboard with Pitch the blade of a
sharp Knife, to intercept some part of the Light
which passed through the hole. The Planes of the
Pasteboard and blade of the Knife were parallel to
one another, and perpendicular to the Rays. And
when they were so placed that none of the Sun's
Light fell on the Pasteboard, but all of it passed
through the hole to the Knife, and there part of it
fell upon the blade of the Knife, and part of it passed
by its edge; I let this part of the Light which passed

by, fall on a white Paper two or three Feet beyond
the Knife, and there saw two streams of faint Light
shoot out both ways from the beam of Light into the
shadow, like the Tails of Comets. But because the
Sun's direct Light by its brightness upon the Paper
obscured these faint streams, so that I could scarce
see them, I made a little hole in the midst of the
Paper for that Light to pass through and fall on a
black Cloth behind it; and then I saw the two
streams plainly. They were like one another, and
pretty nearly equal in length, and breadth, and
quantity of Light. Their Light at that end next the
Sun's direct Light was pretty strong for the space of
about a quarter of an Inch, or half an Inch, and in all
its progress from that direct Light decreased gradually till it became insensible. The whole length of
either of these streams measured upon the paper at
the distance of three Feet from the Knife was about
six or eight Inches; so that it subtended an Angle at
the edge of the Knife of about 10 or 12, or at most
14 Degrees. Yet sometimes I thought I saw it shoot
three or four Degrees farther, but with a Light so
very faint that I could scarce perceive it, and suspected it might (in some measure at least) arise from
some other cause than the two streams did. For placing my Eye in that Light beyond the end of that
stream which was behind the Knife, and looking
towards the Knife, I could see a line of Light upon
its edge, and that not only when my Eye was in the
line of the Streams, but also when it was without
that line either towards the point of the Knife, or

towards the handle. This line of Light appear'd contiguous to the edge of the Knife, and was narrower than the Light of the innermost Fringe, and narrowest when my Eye was farthest from the direct Light, and therefore seem'd to pass between the Light of that Fringe and the edge of the Knife, and that which passed nearest the edge to be most bent, though not all of it.

Obs. 6. I placed another Knife by this, so that their edges might be parallel, and look towards one another, and that the beam of Light might fall upon both the Knives, and some part of it pass between their edges. And when the distance of their edges was about the 400th part of an Inch, the stream parted in the middle, and left a Shadow between the two parts. This Shadow was so black and dark that all the Light which passed between the Knives seem'd to be bent, and turn'd aside to the one hand or to the other. And as the Knives still approach'd one another the Shadow grew broader, and the streams shorter at their inward ends which were next the Shadow, until upon the contact of the Knives the whole Light vanish'd, leaving its place to the Shadow.

And hence I gather that the Light which is least bent, and goes to the inward ends of the streams, passes by the edges of the Knives at the greatest distance, and this distance when the Shadow begins to appear between the streams, is about the 800th part of an Inch. And the Light which passes by the edges of the Knives at distances still less and less, is more and more bent, and goes to those parts of the

streams which are farther and farther from the direct
Light; because when the Knives approach one
another till they touch, those parts of the streams
vanish last which are farthest from the direct Light.

Obs. 7. In the fifth Observation the Fringes did
not appear, but by reason of the breadth of the hole
in the Window became so broad as to run into one
another, and by joining, to make one continued Light
in the beginning of the streams. But in the sixth, as
the Knives approached one another, a little before
the Shadow appeared between the two streams, the
Fringes began to appear on the inner ends of the
Streams on either side of the direct Light; three on
one side made by the edge of one Knife, and three
on the other side made by the edge of the other
Knife. They were distinctest when the Knives were
placed at the greatest distance from the hole in the
Window, and still became more distinct by making
the hole less, insomuch that I could sometimes see a
faint lineament of a fourth Fringe beyond the three
above mention'd. And as the Knives continually approach'd one another, the Fringes grew distincter
and larger, until they vanish'd. The outmost Fringe
vanish'd first, and the middlemost next, and the
innermost last. And after they were all vanish'd, and
the line of Light which was in the middle between
them was grown very broad, enlarging it self on both
sides into the streams of Light described in the fifth
Observation, the above-mention'd Shadow began to
appear in the middle of this line, and divide it along
the middle into two lines of Light, and increased

until the whole Light vanish'd. This enlargement of the Fringes was so great that the Rays which go to the innermost Fringe seem'd to be bent above twenty times more when this Fringe was ready to vanish, than when one of the Knives was taken away.

And from this and the former Observation compared, I gather, that the Light of the first Fringe passed by the edge of the Knife at a distance greater than the 800th part of an Inch, and the Light of the second Fringe passed by the edge of the Knife at a greater distance than the Light of the first Fringe did, and that of the third at a greater distance than that of the second, and that of the streams of Light described in the fifth and sixth Observations passed by the edges of the Knives at less distances than that of any of the Fringes.

Obs. 8. I caused the edges of two Knives to be ground truly strait, and pricking their points into a Board so that their edges might look towards one another, and meeting near their points contain a rectilinear Angle, I fasten'd their Handles together with Pitch to make this Angle invariable. The distance of the edges of the Knives from one another at the distance of four Inches from the angular Point, where the edges of the Knives met, was the eighth part of an Inch; and therefore the Angle contain'd by the edges was about one Degree 54: The Knives thus fix'd together I placed in a beam of the Sun's Light, let into my darken'd Chamber through a Hole the 42d Part of an Inch wide, at the distance of 10 or 15 Feet from the Hole, and let the Light which passed between their edges fall very obliquely upon a smooth white Ruler at the distance of half an Inch, or an Inch from the Knives, and there saw the Fringes by the two edges of the Knives run along the edges of the Shadows of the Knives in Lines parallel to those edges without growing sensibly broader, till they met in Angles equal to the Angle contained by the edges of the Knives, and where they met and joined they ended without crossing one another. But if the Ruler was held at a much greater distance from the Knives, the Fringes where they were farther from the Place of their Meeting, were a little narrower, and became something broader and broader as they approach'd nearer and nearer to one another, and after they met they cross'd one another, and then became much broader than before.

Whence I gather that the distances at which the Fringes pass by the Knives are not increased nor alter'd by the approach of the Knives, but the Angles in which the Rays are there bent are much increased by that approach; and that the Knife which is nearest any Ray determines which way the Ray shall be bent, and the other Knife increases the bent.

Obs. 9. When the Rays fell very obliquely upon the Ruler at the distance of the third Part of an Inch from the Knives, the dark Line between the first and second Fringe of the Shadow of one Knife, and the dark Line between the first and second Fringe of the Shadow of the other knife met with one another, at the distance of the fifth Part of an Inch from the end of the Light which passed between the Knives at the concourse of their edges. And therefore the distance of the edges of the Knives at the meeting of these dark Lines was the 160th Part of an Inch. For as four Inches to the eighth Part of an Inch, so is any Length of the edges of the Knives measured from the point of their concourse to the distance of the edges of the Knives at the end of that Length, and so is the fifth Part of an Inch to the 160th Part. So then the dark Lines above-mention'd meet in the middle of the Light which passes between the Knives where they are distant the 160th Part of an Inch, and the one half of that Light passes by the edge of one Knife at a distance not greater than the 320th Part of an Inch, and falling upon the Paper makes the Fringes of the Shadow of that Knife, and the other half passes by the edge of the other Knife, at a distance not greater than the 320th Part of an Inch, and falling upon the Paper makes the Fringes of the Shadow of the other Knife. But if the Paper be held at a distance from the Knives greater than the third Part of an Inch, the dark Lines above-mention'd meet at a greater distance than the fifth Part of an Inch from the end of the Light which passed between the Knives at the concourse of their edges; and therefore the Light which falls upon the Paper where those dark Lines meet passes between the Knives where the edges are distant above the 160th part of an Inch.

For at another time, when the two Knives were distant eight Feet and five Inches from the little hole in the Window, made with a small Pin as above, the

Light which fell upon the Paper where the aforesaid
dark lines met, passed between the Knives, where
the distance between their edges was as in the following Table, when the distance of the Paper from the
Knives was also as follows.

Distances of the Paper from the Knives in Inches.	Distances between the edges of the Knives in millesimal parts of an Inch.
$1\frac{1}{2}$.	0'012
$3\frac{1}{3}$.	0'020
$8\frac{3}{5}$.	0'034
32.	0'057
96.	0'081
131.	0'087

And hence I gather, that the Light which makes
the Fringes upon the Paper is not the same Light at
all distances of the Paper from the Knives, but when
the Paper is held near the Knives, the Fringes are
made by Light which passes by the edges of the
Knives at a less distance, and is more bent than when
the Paper is held at a greater distance from the
Knives.

Obs. 10. When the Fringes of the Shadows of the
Knives fell perpendicularly upon a Paper at a great
distance from the Knives, they were in the form of
Hyperbola's, and their Dimensions were as follows.
Let CA, CB [in *Fig.* 3.] represent Lines drawn upon
the Paper parallel to the edges of the Knives, and
between which all the Light would fall, if it passed
between the edges of the Knives without inflexion;

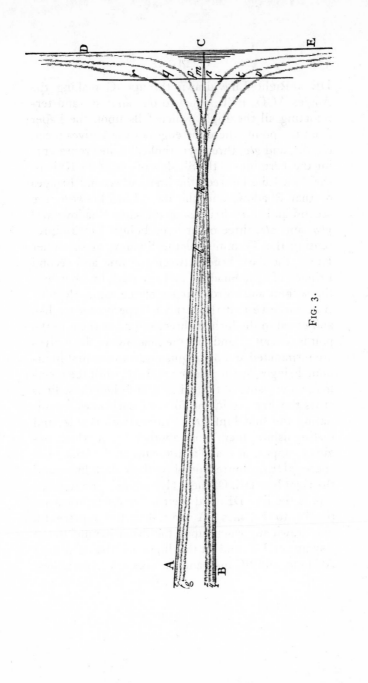

FIG. 3.

DE a Right Line drawn through C making the Angles ACD, BCE, equal to one another, and terminating all the Light which falls upon the Paper from the point where the edges of the Knives meet; *eis*, *fkt*, and *glv*, three hyperbolical Lines representing the Terminus of the Shadow of one of the Knives, the dark Line between the first and second Fringes of that Shadow, and the dark Line between the second and third Fringes of the same Shadow; *xip*, *ykq*, and *zlr*, three other hyperbolical Lines representing the Terminus of the Shadow of the other Knife, the dark Line between the first and second Fringes of that Shadow, and the dark line between the second and third Fringes of the same Shadow. And conceive that these three Hyperbola's are like and equal to the former three, and cross them in the points *i*, *k*, and *l*, and that the Shadows of the Knives are terminated and distinguish'd from the first luminous Fringes by the lines *eis* and *xip*, until the meeting and crossing of the Fringes, and then those lines cross the Fringes in the form of dark lines, terminating the first luminous Fringes within side, and distinguishing them from another Light which begins to appear at *i*, and illuminates all the triangular space *ip*DE*s* comprehended by these dark lines, and the right line DE. Of these Hyperbola's one Asymptote is the line DE, and their other Asymptotes are parallel to the lines CA and CB. Let *rv* represent a line drawn any where upon the Paper parallel to the Asymptote DE, and let this line cross the right lines AC in *m*, and BC in *n*, and the six dark hyperbolical lines in p, q, r; s, t, v; and by measuring the distances ps, qt, rv, and thence collecting the lengths of the Ordinates np, nq, nr or ms, mt, mv, and doing this at several distances of the line rv from the Asymptote DD, you may find as many points of these Hyperbola's as you please, and thereby know that these curve lines are Hyperbola's differing little from the conical Hyperbola. And by measuring the lines Ci, Ck, Cl, you may find other points of these Curves.

For instance; when the Knives were distant from the hole in the Window ten Feet, and the Paper from the Knives nine Feet, and the Angle contained by the edges of the Knives to which the Angle ACB is equal, was subtended by a Chord which was to the Radius as 1 to 32, and the distance of the line rv from the Asymptote DE was half an Inch: I measured the lines ps, qt, rv, and found them 0'35, 0'65, 0'98 Inches respectively; and by adding to their halfs the line $\frac{1}{2}$ mn, (which here was the 128th part of an Inch, or 0'0078 Inches,) the Sums np, nq, nr, were 0'1828, 0'3328, 0'4978 Inches. I measured also the distances of the brightest parts of the Fringes which run between pq and st, qr and tv, and next beyond r and v, and found them 0'5, 0'8, and 1'17 Inches.

Obs. 11. The Sun shining into my darken'd Room through a small round hole made in a Plate of Lead with a slender Pin, as above; I placed at the hole a Prism to refract the Light, and form on the opposite Wall the Spectrum of Colours, described in the third Experiment of the first Book. And then I found that the Shadows of all Bodies held in the colour'd

Light between the Prism and the Wall, were border'd
with Fringes of the Colour of that Light in which
they were held. In the full red Light they were
totally red without any sensible blue or violet, and in
the deep blue Light they were totally blue without
any sensible red or yellow; and so in the green Light
they were totally green, excepting a little yellow
and blue, which were mixed in the green Light of the
Prism. And comparing the Fringes made in the
several colour'd Lights, I found that those made in
the red Light were largest, those made in the violet
were least, and those made in the green were of a
middle bigness. For the Fringes with which the
Shadow of a Man's Hair were bordered, being
measured cross the Shadow at the distance of six
Inches from the Hair, the distance between the
middle and most luminous part of the first or innermost Fringe on one side of the Shadow, and that of
the like Fringe on the other side of the Shadow, was
in the full red Light $\frac{1}{37\frac{1}{4}}$ of an Inch, and in the full
violet $\frac{7}{46}$. And the like distance between the middle
and most luminous parts of the second Fringes on
either side the Shadow was in the full red Light $\frac{1}{22}$,
and in the violet $\frac{1}{27}$ of an Inch. And these distances
of the Fringes held the same proportion at all distances from the Hair without any sensible variation.

So then the Rays which made these Fringes in the
red Light passed by the Hair at a greater distance
than those did which made the like Fringes in the
violet; and therefore the Hair in causing these

Fringes acted alike upon the red Light or least refrangible Rays at a greater distance, and upon the
violet or most refrangible Rays at a less distance, and
by those actions disposed the red Light into Larger
Fringes, and the violet into smaller, and the Lights
of intermediate Colours into Fringes of intermediate
bignesses without changing the Colour of any sort
of Light.

When therefore the Hair in the first and second
of these Observations was held in the white beam of
the Sun's Light, and cast a Shadow which was
border'd with three Fringes of coloured Light, those
Colours arose not from any new modifications impress'd upon the Rays of Light by the Hair, but only
from the various inflexions whereby the several Sorts
of Rays were separated from one another, which
before separation, by the mixture of all their
Colours, composed the white beam of the Sun's
Light, but whenever separated compose Lights of the
several Colours which they are originally disposed
to exhibit. In this 11th Observation, where the
Colours are separated before the Light passes by the
Hair, the least refrangible Rays, which when separated from the rest make red, were inflected at a
greater distance from the Hair, so as to make three
red Fringes at a greater distance from the middle of
the Shadow of the Hair; and the most refrangible
Rays which when separated make violet, were inflected at a less distance from the Hair, so as to
make three violet Fringes at a less distance from the
middle of the Shadow of the Hair. And other Rays

of intermediate degrees of Refrangibility were inflected at intermediate distances from the Hair, so as to make Fringes of intermediate Colours at intermediate distances from the middle of the Shadow of the Hair. And in the second Observation, where all the Colours are mix'd in the white Light which passes by the Hair, these Colours are separated by the various inflexions of the Rays, and the Fringes which they make appear all together, and the innermost Fringes being contiguous make one broad Fringe composed of all the Colours in due order, the violet lying on the inside of the Fringe next the Shadow, the red on the outside farthest from the Shadow, and the blue, green, and yellow, in the middle. And, in like manner, the middlemost Fringes of all the Colours lying in order, and being contiguous, make another broad Fringe composed of all the Colours; and the outmost Fringes of all the Colours lying in order, and being contiguous, make a third broad Fringe composed of all the Colours. These are the three Fringes of colour'd Light with which the Shadows of all Bodies are border'd in the second Observation.

When I made the foregoing Observations, I design'd to repeat most of them with more care and exactness, and to make some new ones for determining the manner how the Rays of Light are bent in their passage by Bodies, for making the Fringes of Colours with the dark lines between them. But I was then interrupted, and cannot now think of taking these things into farther Consideration. And since I

NEWTON IN 1712, FIFTEEN YEARS
BEFORE HIS DEATH

(FROM THE FAMOUS G. KNELLER PORTRAIT)

The FIRST BOOK

OF

OPTICKS.

PART I.

MY Defign in this Book is not to explain the Properties of Light by Hypothefes, but to propofe and prove them by Reafon and Experiments: In order to which, I fhall premife the following Definitions and Axioms.

DEFINITIONS.

DEFIN. I.

BY the Rays of Light I underftand its leaft Parts, and thofe as well Succeffive in the fame Lines as Contemporary in feveral Lines. For it is manifeft that Light confifts of parts both Succeffive and Contemporary; becaufe in the fame place you may ftop that which comes one moment, and let pafs that which comes prefently after; and in the fame time you may ftop it in any one place, and let it pafs in any other. For that part of Light which is ftopt cannot be the fame with that which is let pafs. The leaft Light or part of Light, which may be ftopt alone without the reft of the Light, or propagated alone, or do or fuffer any

A   thing

BEGINNING OF THE TEXT OF THE

1704 EDITION

have not finish'd this part of my Design, I shall conclude with proposing only some Queries, in order to
a farther search to be made by others.

Query 1. Do not Bodies act upon Light at a distance, and by their action bend its Rays; and is
not this action (*cæteris paribus*) strongest at the least
distance?

Qu. 2. Do not the Rays which differ in Refrangibility differ also in Flexibity; and are they not by
their different Inflexions separated from one another,
so as after separation to make the Colours in the three
Fringes above described? And after what manner are
they inflected to make those Fringes?

Qu. 3. Are not the Rays of Light in passing by the
edges and sides of Bodies, bent several times backwards and forwards, with a motion like that of an
Eel? And do not the three Fringes of colour'd Light
above-mention'd arise from three such bendings?

Qu. 4. Do not the Rays of Light which fall upon
Bodies, and are reflected or refracted, begin to bend
before they arrive at the Bodies; and are they not reflected, refracted, and inflected, by one and the same
Principle, acting variously in various Circumstances?

Qu. 5. Do not Bodies and Light act mutually upon
one another; that is to say, Bodies upon Light in
emitting, reflecting, refracting and inflecting it, and
Light upon Bodies for heating them, and putting
their parts into a vibrating motion wherein heat
consists?

Qu. 6. Do not black Bodies conceive heat more
easily from Light than those of other Colours do, by

reason that the Light falling on them is not reflected outwards, but enters the Bodies, and is often reflected and refracted within them, until it be stifled and lost?

Qu. 7. Is not the strength and vigor of the action between Light and sulphureous Bodies observed above, one reason why sulphureous Bodies take fire more readily, and burn more vehemently than other Bodies do?

Qu. 8. Do not all fix'd Bodies, when heated beyond a certain degree, emit Light and shine; and is not this Emission perform'd by the vibrating motions of their parts? And do not all Bodies which abound with terrestrial parts, and especially with sulphureous ones, emit Light as often as those parts are sufficiently agitated; whether that agitation be made by Heat, or by Friction, or Percussion, or Putrefaction, or by any vital Motion, or any other Cause? As for instance; Sea-Water in a raging Storm; Quick-silver agitated in *vacuo*; the Back of a Cat, or Neck of a Horse, obliquely struck or rubbed in a dark place; Wood, Flesh and Fish while they putrefy; Vapours arising from putrefy'd Waters, usually call'd *Ignes Fatui*; Stacks of moist Hay or Corn growing hot by fermentation; Glow-worms and the Eyes of some Animals by vital Motions; the vulgar *Phosphorus* agitated by the attrition of any Body, or by the acid Particles of the Air; Amber and some Diamonds by striking, pressing or rubbing them; Scrapings of Steel struck off with a Flint; Iron hammer'd very nimbly till it become so hot as to kindle Sulphur thrown upon it; the Axletrees of Chariots taking fire by the rapid rotation of the Wheels; and some Liquors mix'd with one another whose Particles come together with an Impetus, as Oil of Vitriol distilled from its weight of Nitre, and then mix'd with twice its weight of Oil of Anniseeds. So also a Globe of Glass about 8 or 10 Inches in diameter, being put into a Frame where it may be swiftly turn'd round its Axis, will in turning shine where it rubs against the palm of ones Hand apply'd to it: And if at the same time a piece of white Paper or white Cloth, or the end of ones Finger be held at the distance of about a quarter of an Inch or half an Inch from that part of the Glass where it is most in motion, the electrick Vapour which is excited by the friction of the Glass against the Hand, will by dashing against the white Paper, Cloth or Finger, be put into such an agitation as to emit Light, and make the white Paper, Cloth or Finger, appear lucid like a Glow-worm; and in rushing out of the Glass will sometimes push against the finger so as to be felt. And the same things have been found by rubbing a long and large Cylinder or Glass or Amber with a Paper held in ones hand, and continuing the friction till the Glass grew warm.

Qu. 9. Is not Fire a Body heated so hot as to emit Light copiously? For what else is a red hot Iron than Fire? And what else is a burning Coal than red hot Wood?

Qu. 10. Is not Flame a Vapour, Fume or Exhalation heated red hot, that is, so hot as to shine? For

Bodies do not flame without emitting a copious
Fume, and this Fume burns in the Flame. The *Ignis
Fatuus* is a Vapour shining without heat, and is there
not the same differénce between this Vapour and
Flame, as between rotten Wood shining without
heat and burning Coals of Fire? In distilling hot
Spirits, if the Head of the Still be taken off, the
Vapour which ascends out of the Still will take fire
at the Flame of a Candle, and turn into Flame, and
the Flame will run along the Vapour from the Candle
to the Still. Some Bodies heated by Motion, or Fermentation, if the heat grow intense, fume copiously,
and if the heat be great enough the Fumes will shine
and become Flame. Metals in fusion do not flame
for want of a copious Fume, except Spelter, which
fumes copiously, and thereby flames. All flaming
Bodies, as Oil, Tallow, Wax, Wood, fossil Coals,
Pitch, Sulphur, by flaming waste and vanish into
burning Smoke, which Smoke, if the Flame be put
out, is very thick and visible, and sometimes smells
strongly, but in the Flame loses its smell by burning,
and according to the nature of the Smoke the Flame
is of several Colours, as that of Sulphur blue, that of
Copper open'd with sublimate green, that of Tallow
yellow, that of Camphire white. Smoke passing
through Flame cannot but grow red hot, and red hot
Smoke can have no other appearance than that of
Flame. When Gun-powder takes fire, it goes away
into Flaming Smoke. For the Charcoal and Sulphur
easily take fire, and set fire to the Nitre, and the Spirit
of the Nitre being thereby rarified into Vapour,

rushes out with Explosion much after the manner
that the Vapour of Water rushes out of an Æolipile;
the Sulphur also being volatile is converted into
Vapour, and augments the Explosion. And the acid
Vapour of the Sulphur (namely that which distils
under a Bell into Oil of Sulphur,) entring violently
into the fix'd Body of the Nitre, sets loose the Spirit
of the Nitre, and excites a great Fermentation, whereby the Heat is farther augmented, and the fix'd Body
of the Nitre is also rarified into Fume, and the Explosion is thereby made more vehement and quick.
For if Salt of Tartar be mix'd with Gun-powder,
and that Mixture be warm'd till it takes fire, the Explosion will be more violent and quick than that of
Gun-powder alone; which cannot proceed from any
other cause than the action of the Vapour of the Gun-
powder upon the Salt of Tartar, whereby that Salt
is rarified. The Explosion of Gun-powder arises
therefore from the violent action whereby all the
Mixture being quickly and vehemently heated, is
rarified and converted into Fume and Vapour: which
Vapour, by the violence of that action, becoming so
hot as to shine, appears in the form of Flame.

Qu. 11. Do not great Bodies conserve their heat
the longest, their parts heating one another, and may
not great dense and fix'd Bodies, when heated beyond
a certain degree, emit Light so copiously, as by the
Emission and Re-action of its Light, and the Reflexions and Refractions of its Rays within its Pores
to grow still hotter, till it comes to a certain period
of heat, such as is that of the Sun? And are not the

Sun and fix'd Stars great Earths vehemently hot, whose heat is conserved by the greatness of the Bodies, and the mutual Action and Reaction between them, and the Light which they emit, and whose parts are kept from fuming away, not only by their fixity, but also by the vast weight and density of the Atmospheres incumbent upon them; and very strongly compressing them, and condensing the Vapours and Exhalations which arise from them? For if Water be made warm in any pellucid Vessel emptied of Air, that Water in the *Vacuum* will bubble and boil as vehemently as it would in the open Air in a Vessel set upon the Fire till it conceives a much greater heat. For the weight of the incumbent Atmosphere keeps down the Vapours, and hinders the Water from boiling, until it grow much hotter than is requisite to make it boil *in vacuo*. Also a mixture of Tin and Lead being put upon a red hot Iron *in vacuo* emits a Fume and Flame, but the same Mixture in the open Air, by reason of the incumbent Atmosphere, does not so much as emit any Fume which can be perceived by Sight. In like manner the great weight of the Atmosphere which lies upon the Globe of the Sun may hinder Bodies there from rising up and going away from the Sun in the form of Vapours and Fumes, unless by means of a far greater heat than that which on the Surface of our Earth would very easily turn them into Vapours and Fumes. And the same great weight may condense those Vapours and Exhalations as soon as they shall at any time begin to ascend from the Sun, and make them

presently fall back again into him, and by that action
increase his Heat much after the manner that in our
Earth the Air increases the Heat of a culinary Fire.
And the same weight may hinder the Globe of the
Sun from being diminish'd, unless by the Emission
of Light, and a very small quantity of Vapours and
Exhalations.

Qu. 12. Do not the Rays of Light in falling upon
the bottom of the Eye excite Vibrations in the
Tunica Retina? Which Vibrations, being propagated
along the solid Fibres of the optick Nerves into the
Brain, cause the Sense of seeing.   For because dense
Bodies conserve their Heat a long time, and the
densest Bodies conserve their Heat the longest, the
Vibrations of their parts are of a lasting nature, and
therefore may be propagated along solid Fibres of
uniform dense Matter to a great distance, for conveying into the Brain the impressions made upon all
the Organs of Sense. For that Motion which can continue long in one and the same part of a Body, can be
propagated a long way from one part to another,
supposing the Body homogeneal, so that the Motion
may not be reflected, refracted, interrupted or dis-
order'd by any unevenness of the Body.

Qu. 13. Do not several sorts of Rays make Vibrations of several bignesses, which according to their
bignesses excite Sensations of several Colours, much
after the manner that the Vibrations of the Air, according to their several bignesses excite Sensations
of several Sounds? And particularly do not the most
refrangible Rays excite the shortest Vibrations for

making a Sensation of deep violet, the least refrangible the largest for making a Sensation of deep red, and the several intermediate sorts of Rays, Vibrations of several intermediate bignesses to make Sensations of the several intermediate Colours?

Qu. 14. May not the harmony and discord of Colours arise from the proportions of the Vibrations propagated through the Fibres of the optick Nerves into the Brain, as the harmony and discord of Sounds arise from the proportions of the Vibrations of the Air? For some Colours, if they be view'd together, are agreeable to one another, as those of Gold and Indigo, and others disagree.

Qu. 15. Are not the Species of Objects seen with both Eyes united where the optick Nerves meet before they come into the Brain, the Fibres on the right side of both Nerves uniting there, and after union going thence into the Brain in the Nerve which is on the right side of the Head, and the Fibres on the left side of both Nerves uniting in the same place, and after union going into the Brain in the Nerve which is on the left side of the Head, and these two Nerves meeting in the Brain in such a manner that their Fibres make but one entire Species or Picture, half of which on the right side of the Sensorium comes from the right side of both Eyes through the right side of both optick Nerves to the place where the Nerves meet, and from thence on the right side of the Head into the Brain, and the other half on the left side of the Sensorium comes in like manner from the left side of both Eyes. For the optick Nerves of such Animals as look the same way with both Eyes (as of Men, Dogs, Sheep, Oxen, &c.) meet before they come into the Brain, but the optick Nerves of such Animals as do not look the same way with both Eyes (as of Fishes, and of the Chameleon,) do not meet, if I am rightly inform'd.

Qu. 16. When a Man in the dark presses either corner of his Eye with his Finger, and turns his Eye away from his Finger, he will see a Circle of Colours like those in the Feather of a Peacock's Tail. If the Eye and the Finger remain quiet these Colours vanish in a second Minute of Time, but if the Finger be moved with a quavering Motion they appear again. Do not these Colours arise from such Motions excited in the bottom of the Eye by the Pressure and Motion of the Finger, as, at other times are excited there by Light for causing Vision? And do not the Motions once excited continue about a Second of Time before they cease? And when a Man by a stroke upon his Eye sees a flash of Light, are not the like Motions excited in the *Retina* by the stroke? And when a Coal of Fire moved nimbly in the circumference of a Circle, makes the whole circumference appear like a Circle of Fire; is it not because the Motions excited in the bottom of the Eye by the Rays of Light are of a lasting nature, and continue till the Coal of Fire in going round returns to its former place? And considering the lastingness of the Motions excited in the bottom of the Eye by Light, are they not of a vibrating nature?

Qu. 17. If a stone be thrown into stagnating Water,

the Waves excited thereby continue some time to
arise in the place where the Stone fell into the Water,
and are propagated from thence in concentrick
Circles upon the Surface of the Water to great distances. And the Vibrations or Tremors excited in
the Air by percussion, continue a little time to move
from the place of percussion in concentrick Spheres
to great distances. And in like manner, when a Ray
of Light falls upon the Surface of any pellucid Body,
and is there refracted or reflected, may not Waves
of Vibrations, or Tremors, be thereby excited in the
refracting or reflecting Medium at the point of Incidence, and continue to arise there, and to be propagated from thence as long as they continue to arise
and be propagated, when they are excited in the
bottom of the Eye by the Pressure or Motion of the
Finger, or by the Light which comes from the Coal
of Fire in the Experiments abovemention'd? and are
not these Vibrations propagated from the point of
Incidence to great distances? And do they not overtake the Rays of Light, and by overtaking them
successively, do they not put them into the Fits of
easy Reflexion and easy Transmission described
above? For if the Rays endeavour to recede from the
densest part of the Vibration, they may be alternately accelerated and retarded by the Vibrations
overtaking them.

Qu. 18. If in two large tall cylindrical Vessels of
Glass inverted, two little Thermometers be suspended so as not to touch the Vessels, and the Air be
drawn out of one of these Vessels, and these Vessels

thus prepared be carried out of a cold place into a warm one; the Thermometer *in vacuo* will grow warm as much, and almost as soon as the Thermometer which is not *in vacuo*. And when the Vessels are carried back into the cold place, the Thermometer *in vacuo* will grow cold almost as soon as the other Thermometer. Is not the Heat of the warm Room convey'd through the *Vacuum* by the Vibrations of a much subtiler Medium than Air, which after the Air was drawn out remained in the *Vacuum*? And is not this Medium the same with that Medium by which Light is refracted and reflected, and by whose Vibrations Light communicates Heat to Bodies, and is put into Fits of easy Reflexion and easy Transmission? And do not the Vibrations of this Medium in hot Bodies contribute to the intenseness and duration of their Heat? And do not hot Bodies communicate their Heat to contiguous cold ones, by the Vibrations of this Medium propagated from them into the cold ones? And is not this Medium exceedingly more rare and subtile than the Air, and exceedingly more elastick and active? And doth it not readily pervade all Bodies? And is it not (by its elastick force) expanded through all the Heavens?

Qu. 19. Doth not the Refraction of Light proceed from the different density of this Æthereal Medium in different places, the Light receding always from the denser parts of the Medium? And is not the density thereof greater in free and open Spaces void of Air and other grosser Bodies, than within the Pores of Water, Glass, Crystal, Gems, and other compact

Bodies? For when Light passes through Glass or
Crystal, and falling very obliquely upon the farther
Surface thereof is totally reflected, the total Reflexion ought to proceed rather from the density and
vigour of the Medium without and beyond the Glass,
than from the rarity and weakness thereof.

Qu. 20. Doth not this Æthereal Medium in passing
out of Water, Glass, Crystal, and other compact and
dense Bodies into empty Spaces, grow denser and
denser by degrees, and by that means refract the
Rays of Light not in a point, but by bending them
gradually in curve Lines? And doth not the gradual
condensation of this Medium extend to some distance from the Bodies, and thereby cause the Inflexions of the Rays of Light, which pass by the
edges of dense Bodies, at some distance from the
Bodies?

Qu. 21. Is not this Medium much rarer within the
dense Bodies of the Sun, Stars, Planets and Comets,
than in the empty celestial Spaces between them?
And in passing from them to great distances, doth it
not grow denser and denser perpetually, and thereby
cause the gravity of those great Bodies towards one
another, and of their parts towards the Bodies; every
Body endeavouring to go from the denser parts of
the Medium towards the rarer? For if this Medium
be rarer within the Sun's Body than at its Surface,
and rarer there than at the hundredth part of an Inch
from its Body, and rarer there than at the fiftieth part
of an Inch from its Body, and rarer there than at the
Orb of *Saturn*; I see no reason why the Increase of

density should stop any where, and not rather be continued through all distances from the Sun to *Saturn*, and beyond. And though this Increase of density may at great distances be exceeding slow, yet if the elastick force of this Medium be exceeding great, it may suffice to impel Bodies from the denser parts of the Medium towards the rarer, with all that power which we call Gravity. And that the elastick force of this Medium is exceeding great, may be gather'd from the swiftness of its Vibrations. Sounds move about 1140 *English* Feet in a second Minute of Time, and in seven or eight Minutes of Time they move about one hundred *English* Miles. Light moves from the Sun to us in about seven or eight Minutes of Time, which distance is about 70000000 *English* Miles, supposing the horizontal Parallax of the Sun to be about 12″. And the Vibrations or Pulses of this Medium, that they may cause the alternate Fits of easy Transmission and easy Reflexion, must be swifter than Light, and by consequence above 700000 times swifter than Sounds. And therefore the elastick force of this Medium, in proportion to its density, must be above 700000 × 700000 (that is, above 490000000000) times greater than the elastick force of the Air is in proportion to its density. For the Velocities of the Pulses of elastick Mediums are in a subduplicate *Ratio* of the Elasticities and the Rarities of the Mediums taken together.

As Attraction is stronger in small Magnets than in great ones in proportion to their Bulk, and Gravity is greater in the Surfaces of small Planets than in those of great ones in proportion to their bulk, and small Bodies are agitated much more by electric attraction than great ones; so the smallness of the Rays of Light may contribute very much to the power of the Agent by which they are refracted. And so if any one should suppose that *Æther* (like our Air) may contain Particles which endeavour to recede from one another (for I do not know what this *Æther* is) and that its Particles are exceedingly smaller than those of Air, or even than those of Light: The exceeding smallness of its Particles may contribute to the greatness of the force by which those Particles may recede from one another, and thereby make that Medium exceedingly more rare and elastick than Air, and by consequence exceedingly less able to resist the motions of Projectiles, and exceedingly more able to press upon gross Bodies, by endeavouring to expand it self.

Qu. 22. May not Planets and Comets, and all gross Bodies, perform their Motions more freely, and with less resistance in this Æthereal Medium than in any Fluid, which fills all Space adequately without leaving any Pores, and by consequence is much denser than Quick-silver or Gold? And may not its resistance be so small, as to be inconsiderable? For instance; If this *Æther* (for so I will call it) should be supposed 700000 times more elastick than our Air, and above 700000 times more rare; its resistance would be above 600000000 times less than that of Water. And so small a resistance would scarce make any sensible alteration in the Motions of the Planets

in ten thousand Years. If any one would ask how a
Medium can be so rare, let him tell me how the Air,
in the upper parts of the Atmosphere, can be above
an hundred thousand thousand times rarer than
Gold. Let him also tell me, how an electrick Body
can by Friction emit an Exhalation so rare and
subtile, and yet so potent, as by its Emission to cause
no sensible Diminution of the weight of the electrick
Body, and to be expanded through a Sphere, whose
Diameter is above two Feet, and yet to be able to
agitate and carry up Leaf Copper, or Leaf Gold, at
the distance of above a Foot from the electrick Body?
And how the Effluvia of a Magnet can be so rare and
subtile, as to pass through a Plate of Glass without
any Resistance or Diminution of their Force, and yet
so potent as to turn a magnetick Needle beyond the
Glass?

    Qu. 23. Is not Vision perform'd chiefly by the Vi-
brations of this Medium, excited in the bottom of
the Eye by the Rays of Light, and propagated
through the solid, pellucid and uniform Capillamenta
of the optick Nerves into the place of Sensation?
And is not Hearing perform'd by the Vibrations
either of this or some other Medium, excited in the
auditory Nerves by the Tremors of the Air, and
propagated through the solid, pellucid and uniform
Capillamenta of those Nerves into the place of Sensation? And so of the other Senses.

    Qu. 24. Is not Animal Motion perform'd by the
Vibrations of this Medium, excited in the Brain by
the power of the Will, and propagated from thence

through the solid, pellucid and uniform Capillamenta of the Nerves into the Muscles, for contracting and dilating them? I suppose that the Capillamenta of the Nerves are each of them solid and uniform, that the vibrating Motion of the Æthereal Medium may be propagated along them from one end to the other uniformly, and without interruption: For Obstructions in the Nerves create Palsies. And that they may be sufficiently uniform, I suppose them to be pellucid when view'd singly, tho' the Reflexions in their cylindrical Surfaces may make the whole Nerve (composed of many Capillamenta) appear opake and white. For opacity arises from reflecting Surfaces, such as may disturb and interrupt the Motions of this Medium.

Qu. 25. Are there not other original Properties of the Rays of Light, besides those already described? An instance of another original Property we have in the Refraction of Island Crystal, described first by *Erasmus Bartholine*, and afterwards more exactly by *Hugenius*, in his Book *De la Lumiere*. This Crystal is a pellucid fissile Stone, clear as Water or Crystal of the Rock, and without Colour; enduring a red Heat without losing its transparency, and in a very strong Heat calcining without Fusion. Steep'd a Day or two in Water, it loses its natural Polish. Being rubb'd on Cloth, it attracts pieces of Straws and other light things, like Ambar or Glass; and with *Aqua fortis* it makes an Ebullition. It seems to be a sort of Talk, and is found in form of an oblique Parallelopiped, with six parallelogram Sides and eight solid Angles.

The obtuse Angles of the Parallelograms are each of them 101 Degrees and 52 Minutes; the acute ones 78 Degrees and 8 Minutes. Two of the solid Angles opposite to one another, as C and E, *See the following Scheme, p. 356.* are compassed each of them with three of these obtuse Angles, and each of the other six with one obtuse and two acute ones. It cleaves easily in planes parallel to any of its Sides, and not in any other Planes. It cleaves with a glossy polite Surface not perfectly plane, but with some little unevenness. It is easily scratch'd, and by reason of its softness it takes a Polish very difficultly. It polishes better upon polish'd Looking-glass than upon Metal, and perhaps better upon Pitch, Leather or Parchment. Afterwards it must be rubb'd with a little Oil or white of an Egg, to fill up its Scratches; whereby it will become very transparent and polite. But for several Experiments, it is not necessary to polish it. If a piece of this crystalline Stone be laid upon a Book, every Letter of the Book seen through it will appear double, by means of a double Refraction. And if any beam of Light falls either perpendicularly, or in any oblique Angle upon any Surface of this Crystal, it becomes divided into two beams by means of the same double Refraction. Which beams are of the same Colour with the incident beam of Light, and seem equal to one another in the quantity of their Light, or very nearly equal. One of these Refractions is perform'd by the usual Rule of Opticks, the Sine of Incidence out of Air into this Crystal being to the Sine of Refraction, as five to three. The

other Refraction, which may be called the unusual
Refraction, is perform'd by the following Rule.

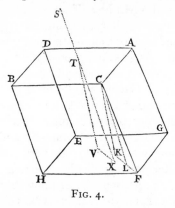

Fig. 4.

Let ADBC represent the refracting Surface of the
Crystal, C the biggest solid Angle at that Surface,
GEHF the opposite Surface, and CK a perpendicular on that Surface. This perpendicular makes
with the edge of the Crystal CF, an Angle of 19
Degr. 3'. Join KF, and in it take KL, so that the
Angle KCL be 6 Degr. 40'. and the Angle LCF 12
Degr. 23'. And if ST represent any beam of Light
incident at T in any Angle upon the refracting Surface ADBC, let TV be the refracted beam determin'd by the given Portion of the Sines 5 to 3, according to the usual Rule of Opticks. Draw VX
parallel and equal to KL. Draw it the same way from
V in which L lieth from K; and joining TX, this line
TX shall be the other refracted beam carried from T
to X, by the unusual Refraction.

If therefore the incident beam ST be perpendicular to the refracting Surface, the two beams TV and TX, into which it shall become divided, shall be parallel to the lines CK and CL; one of those beams going through the Crystal perpendicularly, as it ought to do by the usual Laws of Opticks, and the other TX by an unusual Refraction diverging from the perpendicular, and making with it an Angle VTX of about $6\frac{2}{3}$ Degrees, as is found by Experience. And hence, the Plane VTX, and such like Planes which are parallel to the Plane CFK, may be called the Planes of perpendicular Refraction. And the Coast towards which the lines KL and VX are drawn, may be call'd the Coast of unusual Refraction.

In like manner Crystal of the Rock has a double Refraction: But the difference of the two Refractions is not so great and manifest as in Island Crystal.

When the beam ST incident on Island Crystal is divided into two beams TV and TX, and these two beams arrive at the farther Surface of the Glass; the beam TV, which was refracted at the first Surface after the usual manner, shall be again refracted entirely after the usual manner at the second Surface; and the beam TX, which was refracted after the unusual manner in the first Surface, shall be again refracted entirely after the unusual manner in the second Surface; so that both these beams shall emerge out of the second Surface in lines parallel to the first incident beam ST.

And if two pieces of Island Crystal be placed one after another, in such manner that all the Surfaces of the latter be parallel to all the corresponding Surfaces of the former: The Rays which are refracted after the usual manner in the first Surface of the first Crystal, shall be refracted after the usual manner in all the following Surfaces; and the Rays which are refracted after the unusual manner in the first Surface, shall be refracted after the unusual manner in all the following Surfaces. And the same thing happens, though the Surfaces of the Crystals be any ways inclined to one another, provided that their Planes of perpendicular Refraction be parallel to one another.

And therefore there is an original difference in the Rays of Light, by means of which some Rays are in this Experiment constantly refracted after the usual manner, and others constantly after the unusual manner: For if the difference be not original, but arises from new Modifications impress'd on the Rays at their first Refraction, it would be alter'd by new Modifications in the three following Refractions; whereas it suffers no alteration, but is constant, and has the same effect upon the Rays in all the Refractions. The unusual Refraction is therefore perform'd by an original property of the Rays. And it remains to be enquired, whether the Rays have not more original Properties than are yet discover'd.

Qu. 26. Have not the Rays of Light several sides, endued with several original Properties? For if the Planes of perpendicular Refraction of the second Crystal be at right Angles with the Planes of perpendicular Refraction of the first Crystal, the Rays which

are refracted after the usual manner in passing
through the first Crystal, will be all of them refracted
after the unusual manner in passing through the
second Crystal; and the Rays which are refracted
after the unusual manner in passing through the
first Crystal, will be all of them refracted after the
usual manner in passing through the second Crystal.
And therefore there are not two sorts of Rays differing in their nature from one another, one of which is
constantly and in all Positions refracted after the
usual manner, and the other constantly and in all
Positions after the unusual manner. The difference
between the two sorts of Rays in the Experiment
mention'd in the 25th Question, was only in the
Positions of the Sides of the Rays to the Planes of
perpendicular Refraction. For one and the same Ray
is here refracted sometimes after the usual, and sometimes after the unusual manner, according to the
Position which its Sides have to the Crystals. If the
Sides of the Ray are posited the same way to both
Crystals, it is refracted after the same manner in them
both: But if that side of the Ray which looks towards
the Coast of the unusual Refraction of the first
Crystal, be 90 Degrees from that side of the same
Ray which looks toward the Coast of the unusual
Refraction of the second Crystal, (which may be
effected by varying the Position of the second
Crystal to the first, and by consequence to the Rays
of Light,) the Ray shall be refracted after several
manners in the several Crystals. There is nothing
more required to determine whether the Rays of Light

which fall upon the second Crystal shall be refracted
after the usual or after the unusual manner, but to
turn about this Crystal, so that the Coast of this
Crystal's unusual Refraction may be on this or on
that side of the Ray. And therefore every Ray may
be consider'd as having four Sides or Quarters, two
of which opposite to one another incline the Ray to
be refracted after the unusual manner, as often as
either of them are turn'd towards the Coast of unusual Refraction; and the other two, whenever either
of them are turn'd towards the Coast of unusual Refraction, do not incline it to be otherwise refracted
than after the usual manner. The two first may therefore be call'd the Sides of unusual Refraction. And
since these Dispositions were in the Rays before their
Incidence on the second, third, and fourth Surfaces
of the two Crystals, and suffered no alteration (so far
as appears,) by the Refraction of the Rays in their
passage through those Surfaces, and the Rays were
refracted by the same Laws in all the four Surfaces;
it appears that those Dispositions were in the Rays
originally, and suffer'd no alteration by the first Refraction, and that by means of those Dispositions
the Rays were refracted at their Incidence on the first
Surface of the first Crystal, some of them after the
usual, and some of them after the unusual manner,
accordingly as their Sides of unusual Refraction
were then turn'd towards the Coast of the unusual
Refraction of that Crystal, or sideways from it.

Every Ray of Light has therefore two opposite
Sides, originally endued with a Property on which the

unusual Refraction depends, and the other two opposite Sides not endued with that Property. And it remains to be enquired, whether there are not more Properties of Light by which the Sides of the Rays differ, and are distinguished from one another.

In explaining the difference of the Sides of the Rays above mention'd, I have supposed that the Rays fall perpendicularly on the first Crystal. But if they fall obliquely on it, the Success is the same. Those Rays which are refracted after the usual manner in the first Crystal, will be refracted after the unusual manner in the second Crystal, supposing the Planes of perpendicular Refraction to be at right Angles with one another, as above; and on the contrary.

If the Planes of the perpendicular Refraction of the two Crystals be neither parallel nor perpendicular to one another, but contain an acute Angle: The two beams of Light which emerge out of the first Crystal, will be each of them divided into two more at their Incidence on the second Crystal. For in this case the Rays in each of the two beams will some of them have their Sides of unusual Refraction, and some of them their other Sides turn'd towards the Coast of the unusual Refraction of the second Crystal.

Qu. 27. Are not all Hypotheses erroneous which have hitherto been invented for explaining the Phænomena of Light, by new Modifications of the Rays? For those Phænomena depend not upon new Modifications, as has been supposed, but upon the original and unchangeable Properties of the Rays.

Qu. 28. Are not all Hypotheses erroneous, in which Light is supposed to consist in Pression or Motion, propagated through a fluid Medium? For in all these Hypotheses the Phænomena of Light have been hitherto explain'd by supposing that they arise from new Modifications of the Rays; which is an erroneous Supposition.

If Light consisted only in Pression propagated without actual Motion, it would not be able to agitate and heat the Bodies which refract and reflect it. If it consisted in Motion propagated to all distances in an instant, it would require an infinite force every moment, in every shining Particle, to generate that Motion. And if it consisted in Pression or Motion, propagated either in an instant or in time, it would bend into the Shadow. For Pression or Motion cannot be propagated in a Fluid in right Lines, beyond an Obstacle which stops part of the Motion, but will bend and spread every way into the quiescent Medium which lies beyond the Obstacle. Gravity tends downwards, but the Pressure of Water arising from Gravity tends every way with equal Force, and is propagated as readily, and with as much force sideways as downwards, and through crooked passages as through strait ones. The Waves on the Surface of stagnating Water, passing by the sides of a broad Obstacle which stops part of them, bend afterwards and dilate themselves gradually into the quiet Water behind the Obstacle. The Waves, Pulses or Vibrations of the Air, wherein Sounds consist, bend manifestly, though not so much as the Waves of Water.

For a Bell or a Cannon may be heard beyond a Hill
which intercepts the sight of the sounding Body, and
Sounds are propagated as readily through crooked
Pipes as through streight ones. But Light is never
known to follow crooked Passages nor to bend into
the Shadow. For the fix'd Stars by the Interposition
of any of the Planets cease to be seen. And so do the
Parts of the Sun by the Interposition of the Moon,
Mercury or *Venus*. The Rays which pass very near to
the edges of any Body, are bent a little by the action
of the Body, as we shew'd above; but this bending is
not towards but from the Shadow, and is perform'd
only in the passage of the Ray by the Body, and at a
very small distance from it. So soon as the Ray is
past the Body, it goes right on.

To explain the unusual Refraction of Island
Crystal by Pression or Motion propagated, has not
hitherto been attempted (to my knowledge) except
by *Huygens*, who for that end supposed two several
vibrating Mediums within that Crystal. But when he
tried the Refractions in two successive pieces of that
Crystal, and found them such as is mention'd above;
he confessed himself at a loss for explaining them.
For Pressions or Motions, propagated from a shining
Body through an uniform Medium, must be on all
sides alike; whereas by those Experiments it appears,
that the Rays of Light have different Properties in
their different Sides. He suspected that the Pulses of
Æther in passing through the first Crystal might receive certain new Modifications, which might determine them to be propagated in this or that Medium

within the second Crystal, according to the Position
of that Crystal. But what
Mais pour dire comment   Modifications those might
cela se fait, je n'ay rien trove   be he could not say, nor
jusqu' ici qui me satisfasse.
C. H. de la lumiere, c. 5,   think of any thing satisfac-
p. 91.   tory in that Point. And if
he had known that the un-
usual Refraction depends not on new Modifications,
but on the original and unchangeable Dispositions
of the Rays, he would have found it as difficult to
explain how those Dispositions which he supposed
to be impress'd on the Rays by the first Crystal,
could be in them before their Incidence on that
Crystal, and in general, how all Rays emitted by
shining Bodies, can have those Dispositions in them
from the beginning. To me, at least, this seems inexplicable, if Light be nothing else than Pression or
Motion propagated through *Æther*.

And it is as difficult to explain by these Hypotheses, how Rays can be alternately in Fits of easy
Reflexion and easy Transmission; unless perhaps one
might suppose that there are in all Space two Æthereal vibrating Mediums, and that the Vibrations of
one of them constitute Light, and the Vibrations
of the other are swifter, and as often as they overtake the Vibrations of the first, put them into those
Fits. But how two *Æthers* can be diffused through
all Space, one of which acts upon the other, and by
consequence is re-acted upon, without retarding,
shattering, dispersing and confounding one anothers
Motions, is inconceivable. And against filling the

Heavens with fluid Mediums, unless they be exceeding rare, a great Objection arises from the regular and very lasting Motions of the Planets and Comets in all manner of Courses through the Heavens. For thence it is manifest, that the Heavens are void of all sensible Resistance, and by consequence of all sensible Matter.

For the resisting Power of fluid Mediums arises partly from the Attrition of the Parts of the Medium, and partly from the *Vis inertiæ* of the Matter. That part of the Resistance of a spherical Body which arises from the Attrition of the Parts of the Medium is very nearly as the Diameter, or, at the most, as the *Factum* of the Diameter, and the Velocity of the spherical Body together. And that part of the Resistance which arises from the *Vis inertiæ* of the Matter, is as the Square of that *Factum*. And by this difference the two sorts of Resistance may be distinguish'd from one another in any Medium; and these being distinguish'd, it will be found that almost all the Resistance of Bodies of a competent Magnitude moving in Air, Water, Quick-silver, and such like Fluids with a competent Velocity, arises from the *Vis inertiæ* of the Parts of the Fluid.

Now that part of the resisting Power of any Medium which arises from the Tenacity, Friction or Attrition of the Parts of the Medium, may be diminish'd by dividing the Matter into smaller Parts, and making the Parts more smooth and slippery: But that part of the Resistance which arises from the *Vis inertiæ*, is proportional to the Density of the Matter,

and cannot be diminish'd by dividing the Matter
into smaller Parts, nor by any other means than by
decreasing the Density of the Medium. And for these
Reasons the Density of fluid Mediums is very nearly
proportional to their Resistance. Liquors which
differ not much in Density, as Water, Spirit of Wine,
Spirit of Turpentine, hot Oil, differ not much in
Resistance. Water is thirteen or fourteen times
lighter than Quick-silver and by consequence thirteen or fourteen times rarer, and its Resistance is
less than that of Quick-silver in the same Proportion,
or thereabouts, as I have found by Experiments made
with Pendulums. The open Air in which we breathe
is eight or nine hundred times lighter than Water,
and by consequence eight or nine hundred times
rarer, and accordingly its Resistance is less than that
of Water in the same Proportion, or thereabouts; as
I have also found by Experiments made with Pendulums. And in thinner Air the Resistance is still less,
and at length, by ratifying the Air, becomes insensible. For small Feathers falling in the open Air meet
with great Resistance, but in a tall Glass well emptied
of Air, they fall as fast as Lead or Gold, as I have
seen tried several times. Whence the Resistance
seems still to decrease in proportion to the Density
of the Fluid. For I do not find by any Experiments,
that Bodies moving in Quick-silver, Water or Air,
meet with any other sensible Resistance than what
arises from the Density and Tenacity of those sensible Fluids, as they would do if the Pores of those
Fluids, and all other Spaces, were filled with a dense

and subtile Fluid. Now if the Resistance in a Vessel well emptied of Air, was but an hundred times less than in the open Air, it would be about a million of times less than in Quick-silver. But it seems to be much less in such a Vessel, and still much less in the Heavens, at the height of three or four hundred Miles from the Earth, or above. For Mr. *Boyle* has shew'd that Air may be rarified above ten thousand times in Vessels of Glass; and the Heavens are much emptier of Air than any *Vacuum* we can make below. For since the Air is compress'd by the Weight of the incumbent Atmosphere, and the Density of Air is proportional to the Force compressing it, it follows by Computation, that at the height of about seven and a half *English* Miles from the Earth, the Air is four times rarer than at the Surface of the Earth; and at the height of 15 Miles it is sixteen times rarer than that at the Surface of the Earth; and at the height of $22\frac{1}{2}$, 30, or 38 Miles, it is respectively 64, 256, or 1024 times rarer, or thereabouts; and at the height of 76, 152, 228 Miles, it is about 1000000, 1000000000000, or 1000000000000000000 times rarer; and so on.

Heat promotes Fluidity very much by diminishing the Tenacity of Bodies. It makes many Bodies fluid which are not fluid in cold, and increases the Fluidity of tenacious Liquids, as of Oil, Balsam, and Honey, and thereby decreases their Resistance. But it decreases not the Resistance of Water considerably, as it would do if any considerable part of the Resistance of Water arose from the Attrition or Tenacity of its Parts. And therefore the Resistance of Water arises principally and almost entirely from the *Vis inertiæ* of its Matter; and by consequence, if the Heavens were as dense as Water, they would not have much less Resistance than Water; if as dense as Quick-silver, they would not have much less Resistance than Quick-silver; if absolutely dense, or full of Matter without any *Vacuum*, let the Matter be never so subtil and fluid, they would have a greater Resistance than Quick-silver. A solid Globe in such a Medium would lose above half its Motion in moving three times the length of its Diameter, and a Globe not solid (such as are the Planets,) would be retarded sooner.   And therefore to make way for the regular and lasting Motions of the Planets and Comets, it's necessary to empty the Heavens of all Matter, except perhaps some very thin Vapours, Steams, or Effluvia, arising from the Atmospheres of the Earth, Planets, and Comets, and from such an exceedingly rare Æthereal Medium as we described above. A dense Fluid can be of no use for explaining the Phæ-nomena of Nature, the Motions of the Planets and Comets being better explain'd without it. It serves only to disturb and retard the Motions of those great Bodies, and make the Frame of Nature languish: And in the Pores of Bodies, it serves only to stop the vibrating Motions of their Parts, wherein their Heat and Activity consists. And as it is of no use, and hinders the Operations of Nature, and makes her languish, so there is no evidence for its Existence, and therefore it ought to be rejected. And if it be rejected, the Hypotheses that Light consists in Pression or Motion, propagated through such a Medium, are rejected with it.

And for rejecting such a Medium, we have the Authority of those the oldest and most celebrated Philosophers of *Greece* and *Phœnicia*, who made a *Vacuum*, and Atoms, and the Gravity of Atoms, the first Principles of their Philosophy; tacitly attributing Gravity to some other Cause than dense Matter. Later Philosophers banish the Consideration of such a Cause out of natural Philosophy, feigning Hypotheses for explaining all things mechanically, and referring other Causes to Metaphysicks: Whereas the main Business of natural Philosophy is to argue from Phænomena without feigning Hypotheses, and to deduce Causes from Effects, till we come to the very first Cause, which certainly is not mechanical; and not only to unfold the Mechanism of the World, but chiefly to resolve these and such like Questions. What is there in places almost empty of Matter, and whence is it that the Sun and Planets gravitate towards one another, without dense Matter between them? Whence is it that Nature doth nothing in vain; and whence arises all that Order and Beauty which we see in the World? To what end are Comets, and whence is it that Planets move all one and the same way in Orbs concentrick, while Comets move all manner of ways in Orbs very excentrick; and what hinders the fix'd Stars from falling upon one another? How came the Bodies of Animals to be contrived with so much Art, and for what ends were their several Parts? Was the Eye contrived without Skill

in Opticks, and the Ear without Knowledge of
Sounds? How do the Motions of the Body follow
from the Will, and whence is the Instinct in Animals?
Is not the Sensory of Animals that place to which the
sensitive Substance is present, and into which the
sensible Species of Things are carried through the
Nerves and Brain, that there they may be perceived
by their immediate presence to that Substance? And
these things being rightly dispatch'd, does it not
appear from Phænomena that there is a Being incorporeal, living, intelligent, omnipresent, who in
infinite Space, as it were in his Sensory, sees the
things themselves intimately, and throughly perceives them, and comprehends them wholly by their
immediate presence to himself: Of which things the
Images only carried through the Organs of Sense
into our little Sensoriums, are there seen and beheld
by that which in us perceives and thinks. And
though every true Step made in this Philosophy
brings us not immediately to the Knowledge of the
first Cause, yet it brings us nearer to it, and on that
account is to be highly valued.

   Qu. 29. Are not the Rays of Light very small
Bodies emitted from shining Substances? For such
Bodies will pass through uniform Mediums in right
Lines without bending into the Shadow, which is the
Nature of the Rays of Light. They will also be capable
of several Properties, and be able to conserve their
Properties unchanged in passing through several
Mediums, which is another Condition of the Rays of
Light. Pellucid Substances act upon the Rays of

Light at a distance in refracting, reflecting, and inflecting them, and the Rays mutually agitate the Parts of those Substances at a distance for heating them; and this Action and Re-action at a distance very much resembles an attractive Force between Bodies. If Refraction be perform'd by Attraction of the Rays, the Sines of Incidence must be to the Sines of Refraction in a given Proportion, as we shew'd in our Principles of Philosophy: And this Rule is true by Experience. The Rays of Light in going out of Glass into a *Vacuum*, are bent towards the Glass; and if they fall too obliquely on the *Vacuum*, they are bent backwards into the Glass, and totally reflected; and this Reflexion cannot be ascribed to the Resistance of an absolute *Vacuum*, but must be caused by the Power of the Glass attracting the Rays at their going out of it into the *Vacuum*, and bringing them back. For if the farther Surface of the Glass be moisten'd with Water or clear Oil, or liquid and clear Honey, the Rays which would otherwise be reflected will go into the Water, Oil, or Honey; and therefore are not reflected before they arrive at the farther Surface of the Glass, and begin to go out of it. If they go out of it into the Water, Oil, or Honey, they go on, because the Attraction of the Glass is almost balanced and rendered ineffectual by the contrary Attraction of the Liquor. But if they go out of it into a *Vacuum* which has no Attraction to balance that of the Glass, the Attraction of the Glass either bends and refracts them, or brings them back and reflects them. And this is still more evident by laying together two Prisms of Glass, or two Object-glasses of very long Telescopes, the one plane, the other a little convex, and so compressing them that they do not fully touch, nor are too far asunder. For the Light which falls upon the farther Surface of the first Glass where the Interval between the Glasses is not above the ten hundred thousandth Part of an Inch, will go through that Surface, and through the Air or *Vacuum* between the Glasses, and enter into the second Glass, as was explain'd in the first, fourth, and eighth Observations of the first Part of the second Book. But, if the second Glass be taken away, the Light which goes out of the second Surface of the first Glass into the Air or *Vacuum*, will not go on forwards, but turns back into the first Glass, and is reflected; and therefore it is drawn back by the Power of the first Glass, there being nothing else to turn it back. Nothing more is requisite for producing all the variety of Colours, and degrees of Refrangibility, than that the Rays of Light be Bodies of different Sizes, the least of which may take violet the weakest and darkest of the Colours, and be more easily diverted by refracting Surfaces from the right Course; and the rest as they are bigger and bigger, may make the stronger and more lucid Colours, blue, green, yellow, and red, and be more and more difficultly diverted. Nothing more is requisite for putting the Rays of Light into Fits of easy Reflexion and easy Transmission, than that they be small Bodies which by their attractive Powers, or some other Force, stir up Vibrations in what they act upon, which Vibra-

tions being swifter than the Rays, overtake them
successively, and agitate them so as by turns to
increase and decrease their Velocities, and thereby
put them into those Fits. And lastly, the unusual
Refraction of Island-Crystal looks very much as if
it were perform'd by some kind of attractive virtue
lodged in certain Sides both of the Rays, and of the
Particles of the Crystal. For were it not for some
kind of Disposition or Virtue lodged in some Sides
of the Particles of the Crystal, and not in their other
Sides, and which inclines and bends the Rays towards the Coast of unusual Refraction, the Rays
which fall perpendicularly on the Crystal, would not
be refracted towards that Coast rather than towards
any other Coast, both at their Incidence and at their
Emergence, so as to emerge perpendicularly by a
contrary Situation of the Coast of unusual Refraction at the second Surface; the Crystal acting upon
the Rays after they have pass'd through it, and are
emerging into the Air; or, if you please, into a
Vacuum. And since the Crystal by this Disposition
or Virtue does not act upon the Rays, unless when
one of their Sides of unusual Refraction looks towards that Coast, this argues a Virtue or Disposition
in those Sides of the Rays, which answers to, and
sympathizes with that Virtue or Disposition of the
Crystal, as the Poles of two Magnets answer to one
another. And as Magnetism may be intended and
remitted, and is found only in the Magnet and in
Iron: So this Virtue of refracting the perpendicular
Rays is greater in Island-Crystal, less in Crystal of

the Rock, and is not yet found in other Bodies. I do not say that this Virtue is magnetical: It seems to be of another kind. I only say, that whatever it be, it's difficult to conceive how the Rays of Light, unless they be Bodies, can have a permanent Virtue in two of their Sides which is not in their other Sides, and this without any regard to their Position to the Space or Medium through which they pass.

What I mean in this Question by a *Vacuum*, and by the Attractions of the Rays of Light towards Glass or Crystal, may be understood by what was said in the 18th, 19th, and 20th Questions.

Quest. 30. Are not gross Bodies and Light convertible into one another, and may not Bodies receive much of their Activity from the Particles of Light which enter their Composition? For all fix'd Bodies being heated emit Light so long as they continue sufficiently hot, and Light mutually stops in Bodies as often as its Rays strike upon their Parts, as we shew'd above. I know no Body less apt to shine than Water; and yet Water by frequent Distillations changes into fix'd Earth, as Mr. *Boyle* has try'd; and then this Earth being enabled to endure a sufficient Heat, shines by Heat like other Bodies.

The changing of Bodies into Light, and Light into Bodies, is very conformable to the Course of Nature, which seems delighted with Transmutations. Water, which is a very fluid tasteless Salt, she changes by Heat into Vapour, which is a sort of Air, and by Cold into Ice, which is a hard, pellucid, brittle, fusible Stone; and this Stone returns into Water by Heat, and Vapour returns into Water by Cold. Earth by Heat becomes Fire, and by Cold returns into Earth. Dense Bodies by Fermentation rarify into several sorts of Air, and this Air by Fermentation, and sometimes without it, returns into dense Bodies. Mercury appears sometimes in the form of a fluid Metal, sometimes in the form of a hard brittle Metal, sometimes in the form of a corrosive pellucid Salt call'd Sublimate, sometimes in the form of a tasteless, pellucid, volatile white Earth, call'd *Mercurius Dulcis*; or in that of a red opake volatile Earth, call'd Cinnaber; or in that of a red or white Precipitate, or in that of a fluid Salt; and in Distillation it turns into Vapour, and being agitated *in Vacuo*, it shines like Fire. And after all these Changes it returns again into its first form of Mercury. Eggs grow from insensible Magnitudes, and change into Animals; Tadpoles into Frogs; and Worms into Flies. All Birds, Beasts and Fishes, Insects, Trees, and other Vegetables, with their several Parts, grow out of Water and watry Tinctures and Salts, and by Putrefaction return again into watry Substances. And Water standing a few Days in the open Air, yields a Tincture, which (like that of Malt) by standing longer yields a Sediment and a Spirit, but before Putrefaction is fit Nourishment for Animals and Vegetables. And among such various and strange Transmutations, why may not Nature change Bodies into Light, and Light into Bodies?

Quest. 31. Have not the small Particles of Bodies certain Powers, Virtues, or Forces, by which they act at a distance, not only upon the Rays of Light for reflecting, refracting, and inflecting them, but also upon one another for producing a great Part of the Phænomena of Nature? For it's well known, that Bodies act one upon another by the Attractions of Gravity, Magnetism, and Electricity; and these Instances shew the Tenor and Course of Nature, and make it not improbable but that there may be more attractive Powers than these. For Nature is very consonant and conformable to her self. How these Attractions may be perform'd, I do not here consider. What I call Attraction may be perform'd by impulse, or by some other means unknown to me. I use that Word here to signify only in general any Force by which Bodies tend towards one another, whatsoever be the Cause. For we must learn from the Phænomena of Nature what Bodies attract one another, and what are the Laws and Properties of the Attraction, before we enquire the Cause by which the Attraction is perform'd. The Attractions of Gravity, Magnetism, and Electricity, reach to very sensible distances, and so have been observed by vulgar Eyes, and there may be others which reach to so small distances as hitherto escape Observation; and perhaps electrical Attraction may reach to such small distances, even without being excited by Friction.

For when Salt of Tartar runs *per Deliquium*, is not this done by an Attraction between the Particles of the Salt of Tartar, and the Particles of the Water which float in the Air in the form of Vapours? And why does not common Salt, or Salt-petre, or Vitriol,

run *per Deliquium*, but for want of such an Attraction? Or why does not Salt of Tartar draw more
Water out of the Air than in a certain Proportion to
its quantity, but for want of an attractive Force after
it is satiated with Water? And whence is it but from
this attractive Power that Water which alone distils
with a gentle luke-warm Heat, will not distil from
Salt of Tartar without a great Heat? And is it not
from the like attractive Power between the Particles
of Oil of Vitriol and the Particles of Water, that Oil
of Vitriol draws to it a good quantity of Water out of
the Air, and after it is satiated draws no more, and in
Distillation lets go the Water very difficultly? And
when Water and Oil of Vitriol poured successively
into the same Vessel grow very hot in the mixing,
does not this Heat argue a great Motion in the Parts
of the Liquors? And does not this Motion argue,
that the Parts of the two Liquors in mixing coalesce
with Violence, and by consequence rush towards one
another with an accelerated Motion? And when *Aqua
fortis*, or Spirit of Vitriol poured upon Filings of Iron
dissolves the Filings with a great Heat and Ebullition, is not this Heat and Ebullition effected by a
violent Motion of the Parts, and does not that
Motion argue that the acid Parts of the Liquor rush
towards the Parts of the Metal with violence, and run
forcibly into its Pores till they get between its outmost Particles, and the main Mass of the Metal, and
surrounding those Particles loosen them from the
main Mass, and set them at liberty to float off into
the Water? And when the acid Particles, which alone

would distil with an easy Heat, will not separate from
the Particles of the Metal without a very violent
Heat, does not this confirm the Attraction between
them?

When Spirit of Vitriol poured upon common Salt
or Salt-petre makes an Ebullition with the Salt, and
unites with it, and in Distillation the Spirit of the
common Salt or Salt-petre comes over much easier
than it would do before, and the acid part of the
Spirit of Vitriol stays behind; does not this argue
that the fix'd Alcaly of the Salt attracts the acid Spirit
of the Vitriol more strongly than its own Spirit, and
not being able to hold them both, lets go its own?
And when Oil of Vitriol is drawn off from its weight
of Nitre, and from both the Ingredients a compound
Spirit of Nitre is distilled, and two parts of this Spirit
are poured on one part of Oil of Cloves or Carraway
Seeds, or of any ponderous Oil of vegetable or animal
Substances, or Oil of Turpentine thicken'd with a
little Balsam of Sulphur, and the Liquors grow so
very hot in mixing, as presently to send up a burning
Flame; does not this very great and sudden Heat
argue that the two Liquors mix with violence, and
that their Parts in mixing run towards one another
with an accelerated Motion, and clash with the
greatest Force? And is it not for the same reason that
well rectified Spirit of Wine poured on the same
compound Spirit flashes; and that the *Pulvis ful-
minans*, composed of Sulphur, Nitre, and Salt of
Tartar, goes off with a more sudden and violent
Explosion than Gun-powder, the acid Spirits of the

Sulphur and Nitre rushing towards one another, and towards the Salt of Tartar, with so great a violence, as by the shock to turn the whole at once into Vapour and Flame? Where the Dissolution is slow, it makes a slow Ebullition and a gentle Heat; and where it is quicker, it makes a greater Ebullition with more heat; and where it is done at once, the Ebullition is contracted into a sudden Blast or violent Explosion, with a heat equal to that of Fire and Flame. So when a Drachm of the above-mention'd compound Spirit of Nitre was poured upon half a Drachm of Oil of Carraway Seeds *in vacuo*, the Mixture immediately made a flash like Gun-powder, and burst the exhausted Receiver, which was a Glass six Inches wide, and eight Inches deep. And even the gross Body of Sulphur powder'd, and with an equal weight of Iron Filings and a little Water made into Paste, acts upon the Iron, and in five or six hours grows too hot to be touch'd, and emits a Flame. And by these Experiments compared with the great quantity of Sulphur with which the Earth abounds, and the warmth of the interior Parts of the Earth, and hot Springs, and burning Mountains, and with Damps, mineral Coruscations, Earthquakes, hot suffocating Exhalations, Hurricanes, and Spouts; we may learn that sulphureous Steams abound in the Bowels of the Earth and ferment with Minerals, and sometimes take fire with a sudden Coruscation and Explosion; and if pent up in subterraneous Caverns, burst the Caverns with a great shaking of the Earth, as in springing of a Mine. And then the Vapour generated by the Explosion, expiring through the Pores of the Earth, feels hot and suffocates, and makes Tempests and Hurricanes, and sometimes causes the Land to slide, or the Sea to boil, and carries up the Water thereof in Drops, which by their weight fall down again in Spouts. Also some sulphureous Steams, at all times when the Earth is dry, ascending into the Air, ferment there with nitrous Acids, and sometimes taking fire cause Lightning and Thunder, and fiery Meteors. For the Air abounds with acid Vapours fit to promote Fermentations, as appears by the rusting of Iron and Copper in it, the kindling of Fire by blowing, and the beating of the Heart by means of Respiration. Now the above-mention'd Motions are so great and violent as to shew that in Fermentations the Particles of Bodies which almost rest, are put into new Motions by a very potent Principle, which acts upon them only when they approach one another, and causes them to meet and clash with great violence, and grow hot with the motion, and dash one another into pieces, and vanish into Air, and Vapour, and Flame.

When Salt of Tartar *per deliquium*, being poured into the Solution of any Metal, precipitates the Metal and makes it fall down to the bottom of the Liquor in the form of Mud: Does not this argue that the acid Particles are attracted more strongly by the Salt of Tartar than by the Metal, and by the stronger Attraction go from the Metal to the Salt of Tartar? And so when a Solution of Iron in *Aqua fortis* dissolves the *Lapis Calaminaris*, and lets go the Iron, or a Solu-

tion of Copper dissolves Iron immersed in it and lets
go the Copper, or a Solution of Silver dissolves
Copper and lets go the Silver, or a Solution of Mercury in *Aqua fortis* being poured upon Iron, Copper,
Tin, or Lead, dissolves the Metal and lets go the
Mercury; does not this argue that the acid Particles
of the *Aqua fortis* are attracted more strongly by the
Lapis Calaminaris than by Iron, and more strongly
by Iron than by Copper, and more strongly by
Copper than by Silver, and more strongly by Iron,
Copper, Tin, and Lead, than by Mercury? And is it
not for the same reason that Iron requires more *Aqua
fortis* to dissolve it than Copper, and Copper more
than the other Metals; and that of all Metals, Iron is
dissolved most easily, and is most apt to rust; and
next after Iron, Copper?

When Oil of Vitriol is mix'd with a little Water, or
is run *per deliquium*, and in Distillation the Water
ascends difficultly, and brings over with it some part
of the Oil of Vitriol in the form of Spirit of Vitriol,
and this Spirit being poured upon Iron, Copper, or
Salt of Tartar, unites with the Body and lets go the
Water; doth not this shew that the acid Spirit is attracted by the Water, and more attracted by the fix'd
Body than by the Water, and therefore lets go the
Water to close with the fix'd Body? And is it not for
the same reason that the Water and acid Spirits
which are mix'd together in Vinegar, *Aqua fortis*, and
Spirit of Salt, cohere and rise together in Distillation; but if the *Menstruum* be poured on Salt of Tartar, or on Lead, or Iron, or any fix'd Body which

it can dissolve, the Acid by a stronger Attraction adheres to the Body, and lets go the Water? And is it not also from a mutual Attraction that the Spirits of Soot and Sea-Salt unite and compose the Particles of Sal-armoniac, which are less volatile than before, because grosser and freer from Water; and that the Particles of Sal-armoniac in Sublimation carry up the Particles of Antimony, which will not sublime alone; and that the Particles of Mercury uniting with the acid Particles of Spirit of Salt compose Mercury sublimate, and with the Particles of Sulphur, compose Cinnaber; and that the Particles of Spirit of Wine and Spirit of Urine well rectified unite, and letting go the Water which dissolved them, compose a consistent Body; and that in subliming Cinnaber from Salt of Tartar, or from quick Lime, the Sulphur by a stronger Attraction of the Salt or Lime lets go the Mercury, and stays with the fix'd Body; and that when Mercury sublimate is sublimed from Antimony, or from Regulus of Antimony, the Spirit of Salt lets go the Mercury, and unites with the antimonial metal which attracts it more strongly, and stays with it till the Heat be great enough to make them both ascend together, and then carries up the Metal with it in the form of a very fusible Salt, called Butter of Antimony, although the Spirit of Salt alone be almost as volatile as Water, and the Antimony alone as fix'd as Lead?

When *Aqua fortis* dissolves Silver and not Gold, and *Aqua regia* dissolves Gold and not Silver, may it not be said that *Aqua fortis* is subtil enough to penetrate Gold as well as Silver, but wants the attractive Force to give it Entrance; and that *Aqua regia* is subtil enough to penetrate Silver as well as Gold, but wants the attractive Force to give it Entrance? For *Aqua regia* is nothing else than *Aqua fortis* mix'd with some Spirit of Salt, or with Sal-armoniac; and even common Salt dissolved in *Aqua fortis*, enables the *Menstruum* to dissolve Gold, though the Salt be a gross Body. When therefore Spirit of Salt precipitates Silver out of *Aqua fortis*, is it not done by attracting and mixing with the *Aqua fortis*, and not attracting, or perhaps repelling Silver? And when Water precipitates Antimony out of the Sublimate of Antimony and Sal-armoniac, or out of Butter of Antimony, is it not done by its dissolving, mixing with, and weakening the Sal-armoniac or Spirit of Salt, and its not attracting, or perhaps repelling the Antimony? And is it not for want of an attractive virtue between the Parts of Water and Oil, of Quick-silver and Antimony, of Lead and Iron, that these Substances do not mix; and by a weak Attraction, that Quick-silver and Copper mix difficultly; and from a strong one, that Quick-silver and Tin, Antimony and Iron, Water and Salts, mix readily? And in general, is it not from the same Principle that Heat congregates homogeneal Bodies, and separates heterogeneal ones?

When Arsenick with Soap gives a Regulus, and with Mercury sublimate a volatile fusible Salt, like Butter of Antimony, doth not this shew that Arsenick, which is a Substance totally volatile, is compounded of fix'd and volatile Parts, strongly cohering by a mutual Attraction, so that the volatile will not ascend without carrying up the fixed? And so, when an equal weight of Spirit of Wine and Oil of Vitriol are digested together, and in Distillation yield two fragrant and volatile Spirits which will not mix with one another, and a fix'd black Earth remains behind; doth not this shew that Oil of Vitriol is composed of volatile and fix'd Parts strongly united by Attraction, so as to ascend together in form of a volatile, acid, fluid Salt, until the Spirit of Wine attracts and separates the volatile Parts from the fixed? And therefore, since Oil of Sulphur *per Campanam* is of the same Nature with Oil of Vitriol, may it not be inferred, that Sulphur is also a mixture of volatile and fix'd Parts so strongly cohering by Attraction, as to ascend together in Sublimation. By dissolving Flowers of Sulphur in Oil of Turpentine, and distilling the Solution, it is found that Sulphur is composed of an inflamable thick Oil or fat Bitumen, an acid Salt, a very fix'd Earth, and a little Metal. The three first were found not much unequal to one another, the fourth in so small a quantity as scarce to be worth considering. The acid Salt dissolved in Water, is the same with Oil of Sulphur *per Campanam*, and abounding much in the Bowels of the Earth, and particularly in Markasites, unites it self to the other Ingredients of the Markasite, which are, Bitumen, Iron, Copper, and Earth, and with them compounds Allum, Vitriol, and Sulphur. With the Earth alone it compounds Allum; with the Metal alone, or Metal and Earth together, it compounds Vitriol; and with the Bitumen and Earth it compounds Sulphur. Whence it comes to pass that Markasites abound with those three Minerals. And is it not from the mutual Attraction of the Ingredients that they stick together for compounding these Minerals, and that the Bitumen carries up the other Ingredients of the Sulphur, which without it would not sublime? And the same Question may be put concerning all, or almost all the gross Bodies in Nature. For all the Parts of Animals and Vegetables are composed of Substances volatile and fix'd, fluid and solid, as appears by their Analysis; and so are Salts and Minerals, so far as Chymists have been hitherto able to examine their Composition.

When Mercury sublimate is re-sublimed with fresh Mercury, and becomes *Mercurius Dulcis*, which is a white tasteless Earth scarce dissolvable in Water, and *Mercurius Dulcis* re-sublimed with Spirit of Salt returns into Mercury sublimate; and when Metals corroded with a little acid turn into rust, which is an Earth tasteless and indissolvable in Water, and this Earth imbibed with more acid becomes a metallick Salt; and when some Stones, as Spar of Lead, dissolved in proper *Menstruums* become Salts; do not these things shew that Salts are dry Earth and watry Acid united by Attraction, and that the Earth will not become a Salt without so much acid as makes it dissolvable in Water? Do not the sharp and pungent Tastes of Acids arise from the strong Attraction whereby the acid Particles rush upon and agitate the

Particles of the Tongue? And when Metals are dissolved in acid *Menstruums*, and the Acids in conjunction with the Metal act after a different manner, so that the Compound has a different Taste much milder than before, and sometimes a sweet one; is it not because the Acids adhere to the metallick Particles, and thereby lose much of their Activity? And if the Acid be in too small a Proportion to make the Compound dissolvable in Water, will it not by adhering strongly to the Metal become unactive and lose its Taste, and the Compound be a tasteless Earth? For such things as are not dissolvable by the Moisture of the Tongue, act not upon the Taste.

As Gravity makes the Sea flow round the denser and weightier Parts of the Globe of the Earth, so the Attraction may make the watry Acid flow round the denser and compacter Particles of Earth for composing the Particles of Salt. For otherwise the Acid would not do the Office of a Medium between the Earth and common Water, for making Salts dissolvable in the Water; nor would Salt of Tartar readily draw off the Acid from dissolved Metals, nor Metals the Acid from Mercury. Now, as in the great Globe of the Earth and Sea, the densest Bodies by their Gravity sink down in Water, and always endeavour to go towards the Center of the Globe; so in Particles of Salt, the densest Matter may always endeavour to approach the Center of the Particle: So that a Particle of Salt may be compared to a Chaos; being dense, hard, dry, and earthy in the Center; and rare, soft, moist, and watry in the Circumference.

And hence it seems to be that Salts are of a lasting
Nature, being scarce destroy'd, unless by drawing
away their watry Parts by violence, or by letting
them soak into the Pores of the central Earth by a
gentle Heat in Putrefaction, until the Earth be dissolved by the Water, and separated into smaller
Particles, which by reason of their Smallness make
the rotten Compound appear of a black Colour.
Hence also it may be, that the Parts of Animals and
Vegetables preserve their several Forms, and assimilate their Nourishment; the soft and moist Nourishment easily changing its Texture by a gentle Heat
and Motion, till it becomes like the dense, hard, dry,
and durable Earth in the Center of each Particle. But
when the Nourishment grows unfit to be assimilated,
or the central Earth grows too feeble to assimilate
it, the Motion ends in Confusion, Putrefaction, and
Death.

If a very small quantity of any Salt or Vitriol be
dissolved in a great quantity of Water, the Particles
of the Salt or Vitriol will not sink to the bottom,
though they be heavier in Specie than the Water, but
will evenly diffuse themselves into all the Water, so
as to make it as saline at the top as at the bottom. And
does not this imply that the Parts of the Salt or
Vitriol recede from one another, and endeavour to
expand themselves, and get as far asunder as the
quantity of Water in which they float, will allow?
And does not this Endeavour imply that they have a
repulsive Force by which they fly from one another,
or at least, that they attract the Water more strongly

than they do one another? For as all things ascend in Water which are less attracted than Water, by the gravitating Power of the Earth; so all the Particles of Salt which float in Water, and are less attracted than Water by any one Particle of Salt, must recede from that Particle, and give way to the more attracted Water.

When any saline Liquor is evaporated to a Cuticle and let cool, the Salt concretes in regular Figures; which argues, that the Particles of the Salt before they concreted, floated in the Liquor at equal distances in rank and file, and by consequence that they acted upon one another by some Power which at equal distances is equal, at unequal distances unequal. For by such a Power they will range themselves uniformly, and without it they will float irregularly, and come together as irregularly. And since the Particles of Island-Crystal act all the same way upon the Rays of Light for causing the unusual Refraction, may it not be supposed that in the Formation of this Crystal, the Particles not only ranged themselves in rank and file for concreting in regular Figures, but also by some kind of polar Virtue turned their homogeneal Sides the same way.

The Parts of all homogeneal hard Bodies which fully touch one another, stick together very strongly. And for explaining how this may be, some have invented hooked Atoms, which is begging the Question; and others tell us that Bodies are glued together by rest, that is, by an occult Quality, or rather by nothing; and others, that they stick together by conspiring Motions, that is, by relative rest amongst themselves. I had rather infer from their Cohesion, that their Particles attract one another by some Force, which in immediate Contact is exceeding strong, at small distances performs the chymical Operations above-mention'd, and reaches not far from the Particles with any sensible Effect.

All Bodies seem to be composed of hard Particles: For otherwise Fluids would not congeal; as Water, Oils, Vinegar, and Spirit or Oil of Vitriol do by freezing; Mercury by Fumes of Lead; Spirit of Nitre and Mercury, by dissolving the Mercury and evaporating the Flegm; Spirit of Wine and Spirit of Urine, by deflegming and mixing them; and Spirit of Urine and Spirit of Salt, by subliming them together to make Sal-armoniac. Even the Rays of Light seem to be hard Bodies; for otherwise they would not retain different Properties in their different Sides. And therefore Hardness may be reckon'd the Property of all uncompounded Matter. At least, this seems to be as evident as the universal Impenetrability of Matter. For all Bodies, so far as Experience reaches, are either hard, or may be harden'd; and we have no other Evidence of universal Impenetrability, besides a large Experience without an experimental Exception. Now if compound Bodies are so very hard as we find some of them to be, and yet are very porous, and consist of Parts which are only laid together; the simple Particles which are void of Pores, and were never yet divided, must be much harder. For such hard Particles being heaped up together, can scarce touch one another in more than a few Points, and therefore must be separable by much less Force than is requisite to break a solid Particle, whose Parts touch in all the Space between them, without any Pores or Interstices to weaken their Cohesion. And how such very hard Particles which are only laid together and touch only in a few Points, can stick together, and that so firmly as they do, without the assistance of something which causes them to be attracted or press'd towards one another, is very difficult to conceive.

The same thing I infer also from the cohering of two polish'd Marbles *in vacuo*, and from the standing of Quick-silver in the Barometer at the height of 50, 60 or 70 Inches, or above, when ever it is well-purged of Air and carefully poured in, so that its Parts be every where contiguous both to one another and to the Glass. The Atmosphere by its weight presses the Quick-silver into the Glass, to the height of 29 or 30 Inches. And some other Agent raises it higher, not by pressing it into the Glass, but by making its Parts stick to the Glass, and to one another. For upon any discontinuation of Parts, made either by Bubbles or by shaking the Glass, the whole Mercury falls down to the height of 29 or 30 Inches.

And of the same kind with these Experiments are those that follow. If two plane polish'd Plates of Glass (suppose two pieces of a polish'd Looking-glass) be laid together, so that their sides be parallel and at a very small distance from one another, and then their lower edges be dipped into Water, the

Water will rise up between them. And the less the distance of the Glasses is, the greater will be the height to which the Water will rise. If the distance be about the hundredth part of an Inch, the Water will rise to the height of about an Inch; and if the distance be greater or less in any Proportion, the height will be reciprocally proportional to the distance very nearly. For the attractive Force of the Glasses is the same, whether the distance between them be greater or less; and the weight of the Water drawn up is the same, if the height of it be reciprocally proportional to the distance of the Glasses. And in like manner, Water ascends between two Marbles polish'd plane, when their polish'd sides are parallel, and at a very little distance from one another, And if slender Pipes of Glass be dipped at one end into stagnating Water, the Water will rise up within the Pipe, and the height to which it rises will be reciprocally proportional to the Diameter of the Cavity of the Pipe, and will equal the height to which it rises between two Planes of Glass, if the Semi-diameter of the Cavity of the Pipe be equal to the distance between the Planes, or thereabouts. And these Experiments succeed after the same manner *in vacuo* as in the open Air, (as hath been tried before the Royal Society,) and therefore are not influenced by the Weight or Pressure of the Atmosphere.

And if a large Pipe of Glass be filled with sifted Ashes well pressed together in the Glass, and one end of the Pipe be dipped into stagnating Water, the Water will rise up slowly in the Ashes, so as in the space of a Week or Fortnight to reach up within the Glass, to the height of 30 or 40 Inches above the stagnating Water. And the Water rises up to this height by the Action only of those Particles of the Ashes which are upon the Surface of the elevated Water; the Particles which are within the Water, attracting or repelling it as much downwards as upwards. And therefore the Action of the Particles is very strong. But the Particles of the Ashes being not so dense and close together as those of Glass, their Action is not so strong as that of Glass, which keeps Quick-silver suspended to the height of 60 or 70 Inches, and therefore acts with a Force which would keep Water suspended to the height of above 60 Feet.

By the same Principle, a Sponge sucks in Water, and the Glands in the Bodies of Animals, according to their several Natures and Dispositions, suck in various Juices from the Blood.

If two plane polish'd Plates of Glass three or four Inches broad, and twenty or twenty five long, be laid one of them parallel to the Horizon, the other upon the first, so as at one of their ends to touch one another, and contain an Angle of about 10 or 15 Minutes, and the same be first moisten'd on their inward sides with a clean Cloth dipp'd into Oil of Oranges or Spirit of Turpentine, and a Drop or two of the Oil or Spirit be let fall upon the lower Glass at the other; so soon as the upper Glass is laid down upon the lower, so as to touch it at one end as above, and to touch the Drop at the other end, making with the lower Glass an Angle of about 10 or 15

Minutes; the Drop will begin to move towards the Concourse of the Glasses, and will continue to move with an accelerated Motion, till it arrives at that Concourse of the Glasses. For the two Glasses attract the Drop, and make it run that way towards which the Attractions incline. And if when the Drop is in motion you lift up that end of the Glasses where they meet, and towards which the Drop moves, the Drop will ascend between the Glasses, and therefore is attracted. And as you lift up the Glasses more and more, the Drop will ascend slower and slower, and at length rest, being then carried downward by its Weight, as much as upwards by the Attraction. And by this means you may know the Force by which the Drop is attracted at all distances from the Concourse of the Glasses.

Now by some Experiments of this kind, (made by Mr. *Hauksbee*) it has been found that the Attraction is almost reciprocally in a duplicate Proportion of the distance of the middle of the Drop from the Concourse of the Glasses, *viz.* reciprocally in a simple Proportion, by reason of the spreading of the Drop, and its touching each Glass in a larger Surface; and again reciprocally in a simple Proportion, by reason of the Attractions growing stronger within the same quantity of attracting Surface. The Attraction therefore within the same quantity of attracting Surface, is reciprocally as the distance between the Glasses. And therefore where the distance is exceeding small, the Attraction must be exceeding great. By the Table in the second Part of the second Book, wherein the

thicknesses of colour'd Plates of Water between two
Glasses are set down, the thickness of the Plate where
it appears very black, is three eighths of the ten
hundred thousandth part of an Inch. And where the
Oil of Oranges between the Glasses is of this thickness, the Attraction collected by the foregoing Rule,
seems to be so strong, as within a Circle of an Inch in
diameter, to suffice to hold up a Weight equal to that
of a Cylinder of Water of an Inch in diameter, and
two or three Furlongs in length. And where it is of a
less thickness the Attraction may be proportionally
greater, and continue to increase, until the thickness
do not exceed that of a single Particle of the Oil.
There are therefore Agents in Nature able to make
the Particles of Bodies stick together by very strong
Attractions. And it is the Business of experimental
Philosophy to find them out.

Now the smallest Particles of Matter may cohere
by the strongest Attractions, and compose bigger
Particles of weaker Virtue; and many of these may
cohere and compose bigger Particles whose Virtue
is still weaker, and so on for divers Successions, until
the Progression end in the biggest Particles on which
the Operations in Chymistry, and the Colours of
natural Bodies depend, and which by cohering compose Bodies of a sensible Magnitude. If the Body is
compact, and bends or yields inward to Pression
without any sliding of its Parts, it is hard and elastick,
returning to its Figure with a Force rising from the
mutual Attraction of its Parts. If the Parts slide upon
one another, the Body is malleable or soft. If they

slip easily, and are of a fit Size to be agitated by Heat, and the Heat is big enough to keep them in Agitation, the Body is fluid; and if it be apt to stick to things, it is humid; and the Drops of every fluid affect a round Figure by the mutual Attraction of their Parts, as the Globe of the Earth and Sea affects a round Figure by the mutual Attraction of its Parts by Gravity.

Since Metals dissolved in Acids attract but a small quantity of the Acid, their attractive Force can reach but to a small distance from them. And as in Algebra, where affirmative Quantities vanish and cease, there negative ones begin; so in Mechanicks, where Attraction ceases, there a repulsive Virtue ought to succeed. And that there is such a Virtue, seems to follow from the Reflexions and Inflexions of the Rays of Light. For the Rays are repelled by Bodies in both these Cases, without the immediate Contact of the reflecting or inflecting Body. It seems also to follow from the Emission of Light; the Ray so soon as it is shaken off from a shining Body by the vibrating Motion of the Parts of the Body, and gets beyond the reach of Attraction, being driven away with exceeding great Velocity. For that Force which is sufficient to turn it back in Reflexion, may be sufficient to emit it. It seems also to follow from the Production of Air and Vapour. The Particles when they are shaken off from Bodies by Heat or Fermentation, so soon as they are beyond the reach of the Attraction of the Body, receding from it, and also from one another with great Strength, and keeping at a dis-

tance, so as sometimes to take up above a Million of
Times more space than they did before in the form
of a dense Body. Which vast Contraction and Expansion seems unintelligible, by feigning the Particles
of Air to be springy and ramous, or rolled up like
Hoops, or by any other means than a repulsive Power.
The Particles of Fluids which do not cohere too
strongly, and are of such a Smallness as renders them
most susceptible of those Agitations which keep
Liquors in a Fluor, are most easily separated and
rarified into Vapour, and in the Language of the
Chymists, they are volatile, rarifying with an easy
Heat, and condensing with Cold. But those which
are grosser, and so less susceptible of Agitation, or
cohere by a stronger Attraction, are not separated
without a stronger Heat, or perhaps not without Fermentation. And these last are the Bodies which
Chymists call fix'd, and being rarified by Fermentation, become true permanent Air; those Particles receding from one another with the greatest Force, and
being most difficultly brought together, which upon
Contact cohere most strongly. And because the Particles of permanent Air are grosser, and arise from
denser Substances than those of Vapours, thence it
is that true Air is more ponderous than Vapour, and
that a moist Atmosphere is lighter than a dry one,
quantity for quantity. From the same repelling
Power it seems to be that Flies walk upon the Water
without wetting their Feet; and that the Object-
glasses of long Telescopes lie upon one another without touching; and that dry Powders are difficultly

made to touch one another so as to stick together,
unless by melting them, or wetting them with Water,
which by exhaling may bring them together; and
that two polish'd Marbles, which by immediate Contact stick together, are difficultly brought so close
together as to stick.

And thus Nature will be very conformable to her
self and very simple, performing all the great
Motions of the heavenly Bodies by the Attraction of
Gravity which intercedes those Bodies, and almost
all the small ones of their Particles by some other
attractive and repelling Powers which intercede the
Particles. The *Vis inertiæ* is a passive Principle by
which Bodies persist in their Motion or Rest, receive
Motion in proportion to the Force impressing it, and
resist as much as they are resisted. By this Principle
alone there never could have been any Motion in the
World. Some other Principle was necessary for putting Bodies into Motion; and now they are in Motion,
some other Principle is necessary for conserving the
Motion. For from the various Composition of two
Motions, 'tis very certain that there is not always the
same quantity of Motion in the World. For if two
Globes joined by a slender Rod, revolve about their
common Center of Gravity with an uniform Motion,
while that Center moves on uniformly in a right Line
drawn in the Plane of their circular Motion; the Sum
of the Motions of the two Globes, as often as the
Globes are in the right Line described by their
common Center of Gravity, will be bigger than the
Sum of their Motions, when they are in a Line per-

pendicular to that right Line. By this Instance it appears that Motion may be got or lost. But by reason of the Tenacity of Fluids, and Attrition of their Parts, and the Weakness of Elasticity in Solids, Motion is much more apt to be lost than got, and is always upon the Decay. For Bodies which are either absolutely hard, or so soft as to be void of Elasticity, will not rebound from one another. Impenetrability makes them only stop. If two equal Bodies meet directly *in vacuo*, they will by the Laws of Motion stop where they meet, and lose all their Motion, and remain in rest, unless they be elastick, and receive new Motion from their Spring. If they have so much Elasticity as suffices to make them re-bound with a quarter, or half, or three quarters of the Force with which they come together, they will lose three quarters, or half, or a quarter of their Motion. And this may be try'd, by letting two equal Pendulums fall against one another from equal heights. If the Pendulums be of Lead or soft Clay, they will lose all or almost all their Motions: If of elastick Bodies they will lose all but what they recover from their Elasticity. If it be said, that they can lose no Motion but what they communicate to other Bodies, the consequence is, that *in vacuo* they can lose no Motion, but when they meet they must go on and penetrate one another's Dimensions. If three equal round Vessels be filled, the one with Water, the other with Oil, the third with molten Pitch, and the Liquors be stirred about alike to give them a vortical Motion; the Pitch by its Tenacity will lose its Motion quickly, the Oil

being less tenacious will keep it longer, and the Water
being less tenacious will keep it longest, but yet will
lose it in a short time. Whence it is easy to understand, that if many contiguous Vortices of molten
Pitch were each of them as large as those which some
suppose to revolve about the Sun and fix'd Stars, yet
these and all their Parts would, by their Tenacity
and Stiffness, communicate their Motion to one
another till they all rested among themselves. Vortices of Oil or Water, or some fluider Matter, might
continue longer in Motion; but unless the Matter
were void of all Tenacity and Attrition of Parts, and
Communication of Motion, (which is not to be supposed,) the Motion would constantly decay. Seeing
therefore the variety of Motion which we find in the
World is always decreasing, there is a necessity of
conserving and recruiting it by active Principles,
such as are the cause of Gravity, by which Planets
and Comets keep their Motions in their Orbs, and
Bodies acquire great Motion in falling; and the cause
of Fermentation, by which the Heart and Blood of
Animals are kept in perpetual Motion and Heat;
the inward Parts of the Earth are constantly warm'd,
and in some places grow very hot; Bodies burn and
shine, Mountains take fire, the Caverns of the Earth
are blown up, and the Sun continues violently hot
and lucid, and warms all things by his Light. For
we meet with very little Motion in the World,
besides what is owing to these active Principles. And
if it were not for these Principles, the Bodies of the
Earth, Planets, Comets, Sun, and all things in them,

would grow cold and freeze, and become inactive
Masses; and all Putrefaction, Generation, Vegeta-
tion and Life would cease, and the Planets and
Comets would not remain in their Orbs.

All these things being consider'd, it seems probable to me, that God in the Beginning form'd
Matter in solid, massy, hard, impenetrable, moveable Particles, of such Sizes and Figures, and with
such other Properties, and in such Proportion to
Space, as most conduced to the End for which he
form'd them; and that these primitive Particles being
Solids, are incomparably harder than any porous
Bodies compounded of them; even so very hard, as
never to wear or break in pieces; no ordinary Power
being able to divide what God himself made one in
the first Creation. While the Particles continue entire, they may compose Bodies of one and the same
Nature and Texture in all Ages: But should they
wear away, or break in pieces, the Nature of Things
depending on them, would be changed. Water and
Earth, composed of old worn Particles and Fragments of Particles, would not be of the same Nature
and Texture now, with Water and Earth composed
of entire Particles in the Beginning. And therefore,
that Nature may be lasting, the Changes of corporeal
Things are to be placed only in the various Separations and new Associations and Motions of these permanent Particles; compound Bodies being apt to
break, not in the midst of solid Particles, but where
those Particles are laid together, and only touch in a
few Points.

It seems to me farther, that these Particles have not only a *Vis inertiæ*, accompanied with such passive Laws of Motion as naturally result from that Force, but also that they are moved by certain active Principles, such as is that of Gravity, and that which causes Fermentation, and the Cohesion of Bodies. These Principles I consider, not as occult Qualities, supposed to result from the specifick Forms of Things, but as general Laws of Nature, by which the Things themselves are form'd; their Truth appearing to us by Phænomena, though their Causes be not yet discover'd. For these are manifest Qualities, and their Causes only are occult. And the *Aristotelians* gave the Name of occult Qualities, not to manifest Qualities, but to such Qualities only as they supposed to lie hid in Bodies, and to be the unknown Causes of manifest Effects: Such as would be the Causes of Gravity, and of magnetick and electrick Attractions, and of Fermentations, if we should suppose that these Forces or Actions arose from Qualities unknown to us, and uncapable of being discovered and made manifest. Such occult Qualities put a stop to the Improvement of natural Philosophy, and therefore of late Years have been rejected. To tell us that every Species of Things is endow'd with an occult specifick Quality by which it acts and produces manifest Effects, is to tell us nothing: But to derive two or three general Principles of Motion from Phænomena, and afterwards to tell us how the Properties and Actions of all corporeal Things follow from those manifest Principles, would be a very great step in

Philosophy, though the Causes of those Principles were not yet discover'd: And therefore I scruple not to propose the Principles of Motion above-mention'd, they being of very general Extent, and leave their Causes to be found out.

Now by the help of these Principles, all material Things seem to have been composed of the hard and solid Particles above-mention'd, variously associated in the first Creation by the Counsel of an intelligent Agent. For it became him who created them to set them in order. And if he did so, it's unphilosophical to seek for any other Origin of the World, or to pretend that it might arise out of a Chaos by the mere Laws of Nature; though being once form'd, it may continue by those Laws for many Ages. For while Comets move in very excentrick Orbs in all manner of Positions, blind Fate could never make all the Planets move one and the same way in Orbs concentrick, some inconsiderable Irregularities excepted, which may have risen from the mutual Actions of Comets and Planets upon one another, and which will be apt to increase, till this System wants a Reformation. Such a wonderful Uniformity in the Planetary System must be allowed the Effect of Choice. And so must the Uniformity in the Bodies of Animals, they having generally a right and a left side shaped alike, and on either side of their Bodies two Legs behind, and either two Arms, or two Legs, or two Wings before upon their Shoulders, and between their Shoulders a Neck running down into a Back-bone, and a Head upon it; and in the Head two

Ears, two Eyes, a Nose, a Mouth, and a Tongue, alike situated. Also the first Contrivance of those very artificial Parts of Animals, the Eyes, Ears, Brain, Muscles, Heart, Lungs, Midriff, Glands, Larynx, Hands, Wings, swimming Bladders, natural Spectacles, and other Organs of Sense and Motion; and the Instinct of Brutes and Insects, can be the effect of nothing else than the Wisdom and Skill of a powerful ever-living Agent, who being in all Places, is more able by his Will to move the Bodies within his boundless uniform Sensorium, and thereby to form and reform the Parts of the Universe, than we are by our Will to move the Parts of our own Bodies. And yet we are not to consider the World as the Body of God, or the several Parts thereof, as the Parts of God. He is an uniform Being, void of Organs, Members or Parts, and they are his Creatures subordinate to him, and subservient to his Will; and he is no more the Soul of them, than the Soul of Man is the Soul of the Species of Things carried through the Organs of Sense into the place of its Sensation, where it perceives them by means of its immediate Presence, without the Intervention of any third thing. The Organs of Sense are not for enabling the Soul to perceive the Species of Things in its Sensorium, but only for conveying them thither; and God has no need of such Organs, he being every where present to the Things themselves. And since Space is divisible *in infinitum*, and Matter is not necessarily in all places, it may be also allow'd that God is able to create Particles of Matter of several Sizes and

Figures, and in several Proportions to Space, and
perhaps of different Densities and Forces, and thereby to vary the Laws of Nature, and make Worlds of
several sorts in several Parts of the Universe. At least,
I see nothing of Contradiction in all this.

As in Mathematicks, so in Natural Philosophy,
the Investigation of difficult Things by the Method
of Analysis, ought ever to precede the Method of
Composition. This Analysis consists in making Experiments and Observations, and in drawing general
Conclusions from them by Induction, and admitting
of no Objections against the Conclusions, but such
as are taken from Experiments, or other certain
Truths. For Hypotheses are not to be regarded in
experimental Philosophy. And although the arguing
from Experiments and Observations by Induction
be no Demonstration of general Conclusions; yet it
is the best way of arguing which the Nature of
Things admits of, and may be looked upon as so
much the stronger, by how much the Induction is
more general. And if no Exception occur from Phænomena, the Conclusion may be pronounced generally. But if at any time afterwards any Exception
shall occur from Experiments, it may then begin to
be pronounced with such Exceptions as occur. By
this way of Analysis we may proceed from Compounds to Ingredients, and from Motions to the
Forces producing them; and in general, from Effects
to their Causes, and from particular Causes to more
general ones, till the Argument end in the most
general. This is the Method of Analysis: And the

Synthesis consists in assuming the Causes discover'd, and establish'd as Principles, and by them explaining the Phænomena proceeding from them, and proving the Explanations.

In the two first Books of these Opticks, I proceeded by this Analysis to discover and prove the original Differences of the Rays of Light in respect of Refrangibility, Reflexibility, and Colour, and their alternate Fits of easy Reflexion and easy Transmission, and the Properties of Bodies, both opake and pellucid, on which their Reflexions and Colours depend. And these Discoveries being proved, may be assumed in the Method of Composition for explaining the Phænomena arising from them: An Instance of which Method I gave in the End of the first Book. In this third Book I have only begun the Analysis of what remains to be discover'd about Light and its Effects upon the Frame of Nature, hinting several things about it, and leaving the Hints to be examin'd and improv'd by the farther Experiments and Observations of such as are inquisitive. And if natural Philosophy in all its Parts, by pursuing this Method, shall at length be perfected, the Bounds of Moral Philosophy will be also enlarged. For so far as we can know by natural Philosophy what is the first Cause, what Power he has over us, and what Benefits we receive from him, so far our Duty towards him, as well as that towards one another, will appear to us by the Light of Nature. And no doubt, if the Worship of false Gods had not blinded the Heathen, their moral Philosophy would have gone farther than to the four Cardinal Virtues; and instead of teaching the Transmigration of Souls, and to worship the Sun and Moon, and dead Heroes, they would have taught us to worship our true Author and Benefactor, as their Ancestors did under the Government of *Noah* and his Sons before they corrupted themselves.

SOME DOVER SCIENCE BOOKS

A CATALOGUE OF SELECTED DOVER BOOKS
IN ALL FIELDS OF INTEREST

AMERICA'S OLD MASTERS, James T. Flexner. Four men emerged unexpectedly from provincial 18th century America to leadership in European art: Benjamin West, J. S. Copley, C. R. Peale, Gilbert Stuart. Brilliant coverage of lives and contributions. Revised, 1967 edition. 69 plates. 365pp. of text.
21806-6 Paperbound $3.00

FIRST FLOWERS OF OUR WILDERNESS: AMERICAN PAINTING, THE COLONIAL PERIOD, James T. Flexner. Painters, and regional painting traditions from earliest Colonial times up to the emergence of Copley, West and Peale Sr., Foster, Gustavus Hesselius, Feke, John Smibert and many anonymous painters in the primitive manner. Engaging presentation, with 162 illustrations. xxii + 368pp.
22180-6 Paperbound $3.50

THE LIGHT OF DISTANT SKIES: AMERICAN PAINTING, 1760-1835, James T. Flexner. The great generation of early American painters goes to Europe to learn and to teach: West, Copley, Gilbert Stuart and others. Allston, Trumbull, Morse; also contemporary American painters—primitives, derivatives, academics—who remained in America. 102 illustrations. xiii + 306pp. 22179-2 Paperbound $3.50

A HISTORY OF THE RISE AND PROGRESS OF THE ARTS OF DESIGN IN THE UNITED STATES, William Dunlap. Much the richest mine of information on early American painters, sculptors, architects, engravers, miniaturists, etc. The only source of information for scores of artists, the major primary source for many others. Unabridged reprint of rare original 1834 edition, with new introduction by James T. Flexner, and 394 new illustrations. Edited by Rita Weiss. 6⅝ x 9⅝.
21695-0, 21696-9, 21697-7 Three volumes, Paperbound $15.00

EPOCHS OF CHINESE AND JAPANESE ART, Ernest F. Fenollosa. From primitive Chinese art to the 20th century, thorough history, explanation of every important art period and form, including Japanese woodcuts; main stress on China and Japan, but Tibet, Korea also included. Still unexcelled for its detailed, rich coverage of cultural background, aesthetic elements, diffusion studies, particularly of the historical period. 2nd, 1913 edition. 242 illustrations. lii + 439pp. of text.
20364-6, 20365-4 Two volumes, Paperbound $6.00

THE GENTLE ART OF MAKING ENEMIES, James A. M. Whistler. Greatest wit of his day deflates Oscar Wilde, Ruskin, Swinburne; strikes back at inane critics, exhibitions, art journalism; aesthetics of impressionist revolution in most striking form. Highly readable classic by great painter. Reproduction of edition designed by Whistler. Introduction by Alfred Werner. xxxvi + 334pp.
21875-9 Paperbound $3.00

THE PRINCIPLES OF PSYCHOLOGY, William James. The famous long course, complete and unabridged. Stream of thought, time perception, memory, experimental methods—these are only some of the concerns of a work that was years ahead of its time and still valid, interesting, useful. 94 figures. Total of xviii + 1391pp.
20381-6, 20382-4 Two volumes, Paperbound $9.00

THE STRANGE STORY OF THE QUANTUM, Banesh Hoffmann. Non-mathematical but thorough explanation of work of Planck, Einstein, Bohr, Pauli, de Broglie, Schrödinger, Heisenberg, Dirac, Feynman, etc. No technical background needed. "Of books attempting such an account, this is the best," Henry Margenau, Yale. 40-page "Postscript 1959." xii + 285pp.
20518-5 Paperbound $3.00

THE RISE OF THE NEW PHYSICS, A. d'Abro. Most thorough explanation in print of central core of mathematical physics, both classical and modern; from Newton to Dirac and Heisenberg. Both history and exposition; philosophy of science, causality, explanations of higher mathematics, analytical mechanics, electromagnetism, thermodynamics, phase rule, special and general relativity, matrices. No higher mathematics needed to follow exposition, though treatment is elementary to intermediate in level. Recommended to serious student who wishes verbal understanding. 97 illustrations. xvii + 982pp.
20003-5, 20004-3 Two volumes, Paperbound$10.00

GREAT IDEAS OF OPERATIONS RESEARCH, Jagjit Singh. Easily followed non-technical explanation of mathematical tools, aims, results: statistics, linear programming, game theory, queueing theory, Monte Carlo simulation, etc. Uses only elementary mathematics. Many case studies, several analyzed in detail. Clarity, breadth make this excellent for specialist in another field who wishes background. 41 figures. x + 228pp.
21886-4 Paperbound $2.50

GREAT IDEAS OF MODERN MATHEMATICS: THEIR NATURE AND USE, Jagjit Singh. Internationally famous expositor, winner of Unesco's Kalinga Award for science popularization explains verbally such topics as differential equations, matrices, groups, sets, transformations, mathematical logic and other important modern mathematics, as well as use in physics, astrophysics, and similar fields. Superb exposition for layman, scientist in other areas. viii + 312pp.
20587-8 Paperbound $2.75

GREAT IDEAS IN INFORMATION THEORY, LANGUAGE AND CYBERNETICS, Jagjit Singh. The analog and digital computers, how they work, how they are like and unlike the human brain, the men who developed them, their future applications, computer terminology. An essential book for today, even for readers with little math. Some mathematical demonstrations included for more advanced readers. 118 figures. Tables. ix + 338pp.
21694-2 Paperbound $2.50

CHANCE, LUCK AND STATISTICS, Horace C. Levinson. Non-mathematical presentation of fundamentals of probability theory and science of statistics and their applications. Games of chance, betting odds, misuse of statistics, normal and skew distributions, birth rates, stock speculation, insurance. Enlarged edition. Formerly "The Science of Chance." xiii + 357pp.
21007-3 Paperbound $2.50

THE RED FAIRY BOOK, Andrew Lang. Lang's color fairy books have long been children's favorites. This volume includes Rapunzel, Jack and the Bean-stalk and 35 other stories, familiar and unfamiliar. 4 plates, 93 illustrations x + 367pp.
21673-X Paperbound $2.50

THE BLUE FAIRY BOOK, Andrew Lang. Lang's tales come from all countries and all times. Here are 37 tales from Grimm, the Arabian Nights, Greek Mythology, and other fascinating sources. 8 plates, 130 illustrations. xi + 390pp.
21437-0 Paperbound $2.75

HOUSEHOLD STORIES BY THE BROTHERS GRIMM. Classic English-language edition of the well-known tales — Rumpelstiltskin, Snow White, Hansel and Gretel, The Twelve Brothers, Faithful John, Rapunzel, Tom Thumb (52 stories in all). Translated into simple, straightforward English by Lucy Crane. Ornamented with head-pieces, vignettes, elaborate decorative initials and a dozen full-page illustrations by Walter Crane. x + 269pp.
21080-4 Paperbound **$2.00**

THE MERRY ADVENTURES OF ROBIN HOOD, Howard Pyle. The finest modern versions of the traditional ballads and tales about the great English outlaw. Howard Pyle's complete prose version, with every word, every illustration of the first edition. Do not confuse this facsimile of the original (1883) with modern editions that change text or illustrations. 23 plates plus many page decorations. xxii + 296pp.
22043-5 Paperbound $2.75

THE STORY OF KING ARTHUR AND HIS KNIGHTS, Howard Pyle. The finest children's version of the life of King Arthur; brilliantly retold by Pyle, with 48 of his most imaginative illustrations. xviii + 313pp. 6⅛ x 9¼.
21445-1 Paperbound $2.50

THE WONDERFUL WIZARD OF OZ, L. Frank Baum. America's finest children's book in facsimile of first edition with all Denslow illustrations in full color. The edition a child should have. Introduction by Martin Gardner. 23 color plates, scores of drawings. iv + 267pp.
20691-2 Paperbound **$3.50**

THE MARVELOUS LAND OF OZ, L. Frank Baum. The second Oz book, every bit as imaginative as the Wizard. The hero is a boy named Tip, but the Scarecrow and the Tin Woodman are back, as is the Oz magic. 16 color plates, 120 drawings by John R. Neill. 287pp.
20692-0 Paperbound $2.50

THE MAGICAL MONARCH OF MO, L. Frank Baum. Remarkable adventures in a land even stranger than Oz. The best of Baum's books not in the Oz series. 15 color plates and dozens of drawings by Frank Verbeck. xviii + 237pp.
21892-9 Paperbound $2.25

THE BAD CHILD'S BOOK OF BEASTS, MORE BEASTS FOR WORSE CHILDREN, A MORAL ALPHABET, Hilaire Belloc. Three complete humor classics in one volume. Be kind to the frog, and do not call him names . . . and 28 other whimsical animals. Familiar favorites and some not so well known. Illustrated by Basil Blackwell. 156pp.
(USO) 20749-8 Paperbound $1.50

MATHEMATICAL PUZZLES FOR BEGINNERS AND ENTHUSIASTS, Geoffrey Mott-Smith. 189 puzzles from easy to difficult—involving arithmetic, logic, algebra, properties of digits, probability, etc.—for enjoyment and mental stimulus. Explanation of mathematical principles behind the puzzles. 135 illustrations. viii + 248pp.

20198-8 Paperbound $2.00

PAPER FOLDING FOR BEGINNERS, William D. Murray and Francis J. Rigney. Easiest book on the market, clearest instructions on making interesting, beautiful origami. Sail boats, cups, roosters, frogs that move legs, bonbon boxes, standing birds, etc. 40 projects; more than 275 diagrams and photographs. 94pp.

20713-7 Paperbound $1.00

TRICKS AND GAMES ON THE POOL TABLE, Fred Herrmann. 79 tricks and games— some solitaires, some for two or more players, some competitive games—to entertain you between formal games. Mystifying shots and throws, unusual caroms, tricks involving such props as cork, coins, a hat, etc. Formerly *Fun on the Pool Table*. 77 figures. 95pp.

21814-7 Paperbound $1.25

HAND SHADOWS TO BE THROWN UPON THE WALL: A SERIES OF NOVEL AND AMUSING FIGURES FORMED BY THE HAND, Henry Bursill. Delightful picturebook from great-grandfather's day shows how to make 18 different hand shadows: a bird that flies, duck that quacks, dog that wags his tail, camel, goose, deer, boy, turtle, etc. Only book of its sort. vi + 33pp. 6½ x 9¼.

21779-5 Paperbound $1.00

WHITTLING AND WOODCARVING, E. J. Tangerman. 18th printing of best book on market. "If you can cut a potato you can carve" toys and puzzles, chains, chessmen, caricatures, masks, frames, woodcut blocks, surface patterns, much more. Information on tools, woods, techniques. Also goes into serious wood sculpture from Middle Ages to present, East and West. 464 photos, figures. x + 293pp.

20965-2 Paperbound $2.50

HISTORY OF PHILOSOPHY, Julián Marías. Possibly the clearest, most easily followed, best planned, most useful one-volume history of philosophy on the market; neither skimpy nor overfull. Full details on system of every major philosopher and dozens of less important thinkers from pre-Socratics up to Existentialism and later. Strong on many European figures usually omitted. Has gone through dozens of editions in Europe. 1966 edition, translated by Stanley Appelbaum and Clarence Strowbridge. xviii + 505pp.

21739-6 Paperbound $3.50

YOGA: A SCIENTIFIC EVALUATION, Kovoor T. Behanan. Scientific but non-technical study of physiological results of yoga exercises; done under auspices of Yale U. Relations to Indian thought, to psychoanalysis, etc. 16 photos. xxiii + 270pp.

20505-3 Paperbound $2.50

Prices subject to change without notice.
Available at your book dealer or write for free catalogue to Dept. GI, Dover Publications, Inc., 180 Varick St., N. Y., N. Y. 10014. Dover publishes more than 150 books each year on science, elementary and advanced mathematics, biology, music, art, literary history, social sciences and other areas.